DISCRETE MATHEMATICS

in the First Two Years

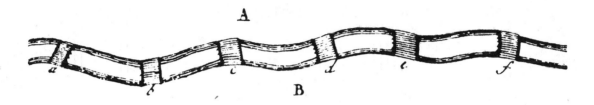

Edited by Anthony Ralston

MAA Notes Series

The MAA Notes Series, started in 1982, addresses a broad range of topics and themes of interest to all who are involved with undergraduate mathematics. The volumes in this series are readable, informative, and useful, and help the mathematical community keep up with developments of importance to mathematics.

1. Problem Solving in the Mathematics Curriculum, Committee on the Undergraduate Teaching of Mathematics, *Alan Schoenfeld,* Editor.
2. Recommendations on the Mathematical Preparation of Teachers, *CUPM Panel on Teacher Training.*
3. Undergraduate Mathematics Education in the People's Republic of China, *Lynn A. Steen,* Editor.
4. Notes on Primality Testing and Factoring, by *Carl Pomerance.*
5. American Perspectives on the Fifth International Congress on Mathematical Education, *Warren Page,* Editor.
6. Toward a Lean and Lively Calculus, *Ronald Douglas,* Editor.
7. Undergraduate Programs in the Mathematical and Computer Sciences: 1985–86, *D. J. Albers, R. D. Anderson, D. O. Loftsgaarden,* Editors.
8. Calculus for a New Century, *Lynn A. Steen,* Editor.
9. Computers and Mathematics: The Use of Computers in Undergraduate Instruction, *D. A. Smith, G. J. Porter, L. C. Leinbach, R. H. Wenger,* Editors.
10. Guidelines for the Continuing Mathematical Education of Teachers, *Committee on the Mathematical Education of Teachers.*
11. Keys to Improved Instruction by Teaching Assistants and Part-Time Instructors, *Committee on Teaching Assistants and Part-Time Instructors, Bettye Anne Case,* Editor.
12. The Use of Calculators in the Standardized Testing of Mathematics, *John Kenelly,* Editor.
13. Reshaping College Mathematics, *Lynn A. Steen,* Editor.
14. Mathematical Writing, by *Donald E. Knuth, Tracy Larrabee,* and *Paul M. Roberts.*
15. Discrete Mathematics in the First Two Years, *Anthony Ralston,* Editor.

First Printing
© 1989 by The Mathematical Association of America
Cover Image: Combinatorics stretches back to antiquity and graph theory reaches back to at least 1735 when Euler wrote his celebrated paper on the problem of Königsberg bridges, from which these illustrations are taken.

Library of Congress number: 89-063055
ISBN 0-88385-064-8
Printed in the United States of America

DISCRETE MATHEMATICS
in the First Two Years

Preface

This volume reports on the experiences of six colleges and universities which, in 1984–86, experimented with various ways of integrating discrete and continuous mathematics in the first two years of college mathematics. The six projects were funded by grants from the Alfred P. Sloan Foundation under a program for which I was a consultant. While this volume was being edited both I and the contributors had opportunity to review the developments since these projects were launched. The changes are remarkable. A single example from my introductory essay gives a sense of these. When these projects were on the drawing boards in 1983 there were almost no appropriate texts available; by 1989 there were about 30 texts available, some in second editions, with more on the way.

The six projects reported on here had varying objectives, used different methods, and met with varying levels of success. As models, they have valuable lessons for those in colleges and universities struggling with how—or whether—to make discrete mathematics a partner of calculus in the freshman and sophomore years. The directors of the six projects have set down narrative discussions, syllabuses, problem sets, and examinations which should be of substantial help to college teachers of discrete mathematics, and, to a lesser extent, to those teaching calculus and linear algebra. The Appendix contains a shortened version of the "Final report of the MAA Committee on discrete mathematics in the first two years." This report has been the point of departure for many efforts to introduce discrete mathematics into the curriculum. In a section preceding this appendix, the chair of the MAA Committee on Discrete Mathematics, Martha J. Siegel, has written some Afterthoughts to the committee's report and she has provided an updated bibliography of discrete mathematics texts.

The project directors and I would welcome comments about the content of this book.

Anthony Ralston
September 1989

Table of Contents

Calculus and Discrete Mathematics - Is the Twain Meeting?

by Anthony Ralston

Much, perhaps most of the American college mathematics community now agrees that discrete mathematics is good and valuable mathematics, particularly appropriate for students of computer science and most of the social and management sciences. Therefore, throughout colleges and universities in the United States departments of mathematics are now teaching discrete mathematics to freshmen and sophomores or they are thinking about doing it. And, they're not just teaching it to computer science students, but to mathematics majors and students in many other disciplines as well, even sometimes engineering and physics. Which is not to say that there are no problems. There are many. What is an appropriate syllabus? One semester or two semesters? Requirement or elective? And what should happen to calculus? This volume does not pretend to answer all - even any - of these questions. But by relating the experience of six institutions which have introduced discrete mathematics into the first two years of their mathematics curriculum, it may give some guidance to others who wish to do the same thing.

Some History

The six colleges and universities whose experiences are reported in this volume were all supported by the Alfred P. Sloan Foundation under a program to foster "the development of a new curriculum for the first two years of undergraduate mathematics in which discrete mathematics [would] play a role of equal importance to that of the calculus." This was the culmination of a program of support by the Sloan Foundation for the development of collegiate discrete mathematics which began in 1978 with a grant to me to study the mathematical needs of computer science students. This study quickly became one of considering the first two years of college mathematics more generally and resulted in the paper [3]. The growing interest in this subject then persuaded the Sloan Foundation to support a conference/workshop at Williams College in the summer of 1982 whose proceedings were reported in [4]. Although by this time a number of colleges and universities were already teaching discrete mathematics in the first two years, it was clear that the problems attendant upon doing this were considerable and that many institutions, interested though they may have been in doing something related to discrete mathematics, were

unsure of what could and should be done and lacked the resources to make the changes necessary to introduce discrete mathematics into the curriculum. What was needed was the support given by the Sloan Foundation for some pilot programs to provide the experience, text materials and problems discussed in this volume. At about the same time the Foundation also funded the work of the MAA Committee on Discrete Mathematics in the First Two Years part of whose report is included in the Appendix to this volume

Calculus and Discrete Mathematics

I take it that it is no longer necessary in writing about discrete mathematics to define what it is. If it is true - and it is - that there is not yet any agreed upon set of topics which should be covered in a one or two semester discrete mathematics course, there is at least general agreement in the mathematics community on what topics fall naturally under the discrete mathematics rubric. For more discussion of this see the Appendix.

The major question that is not yet resolved about the position of discrete mathematics in the college curriculum is: What should be its position in the curriculum vis á vis calculus? You will find a variety of answers and non-answers to this question in the reports of the six schools. One - Florida State - concentrated on the development of a one-year discrete mathematics sequence while leaving the traditional calculus sequence essentially in place. Another - St. Olaf - did essentially the same because of a belief that the way to handle discrete mathematics in the curriculum at this time was not to deemphasize calculus in any way. However, St. Olaf did develop a new course - Applied Calculus - available to students between the first and second semesters of calculus which emphasizes the discrete and numerical aspects of calculus. Moreover, St. Olaf has endeavored to introduce a discrete and numerical flavor into calculus wherever possible. At Montclair State the original plan was to start both mathematics and computer science students with a year of discrete mathematics followed by a year of calculus. But for a variety of reasons the requirement of a year of discrete mathematics for computer science students has been abandoned. Instead computer science students will be required to take a year of calculus and one semester each of probability and linear algebra. Since the latter two courses are arguably both discrete mathematics, the result at Montclair State has been to introduce a considerable amount of discrete mathematics into the first two years for computer science students at least.

At the other three schools serious attempts were made to create a balance between calculus and discrete mathematics in the first two years. At Delaware the four semesters had the structure

Discrete Mathematics – Calculus 1 –
Calculus 2 – Analysis of Algorithms

with the latter being essentially a second semester of discrete mathematics. The key points to note here are

- the inclusion of about 30% more material in Calculus 1 than previously; this seems to have been successful, perhaps because the semester of discrete mathematics gave the students the maturity to handle more material in the calculus

- the movement of some traditional calculus topics like sequences and series to the discrete mathematics courses; this seems sensible because sequences and a considerable portion of series are bona fide discrete mathematics.

The jury on this experiment is still out. Problems in the second semester of calculus and other pressures may result in a return to the old three-semester sequence. It is worth noting that this experiment was undertaken in a large department by only a couple of people with the full support of the department head but without any general agreement in the department that this was a good direction in which to go.

At Denver the calculus sequence was taught in three quarters out of six over two years. One of the other three quarters was devoted to discrete mathematics and another to combinatorics and computational linear algebra. The remaining quarter course combined linear algebra and differential equations with a traditional vector space approach in the linear algebra portion. Some points worth noting are

- as with the Delaware experiment, this was a true attempt to provide a balance between calculus and discrete mathematics.

- the place of linear algebra in such a balance remains to be determined but since so much of linear algebra has a discrete flavor, it is likely that a one-year sequence in discrete mathematics will eventually include a considerable amount of linear algebra, albeit of a more computational, less vector space approach than has been traditional.

Another aspect of the Denver experiment worth noting is a comparison made at the end of the first year (which included two quarters of calculus) with physics and engineering students who had taken a traditional one-year calculus course. The experimental group performed somewhat more poorly on a standardized test but there were some extenuating circumstances.

At Colby the main focus of the Sloan program was on the development of a one-year calculus to go with a later-to-be-developed one-year discrete mathematics sequence. This was successfully done and a book incorporating the one-year calculus syllabus has been written. Thus, Colby, too, has attempted to provide a two-year sequence which balances discrete mathematics and the calculus. Still, it is only fair to note that the Colby sequence is intended only for students who have had some exposure to calculus in high school and, thus, it is reasonable to ask how generally applicable the program which has been developed would be.

What conclusions, if any, can be drawn from all this? Here are mine:

- it is hardly news that for political and inertial reasons, to say nothing of intellectual ones, it will be extraordinarily difficult to create a two-year calculus - discrete mathematics sequence which will satisfy the American mathematical community; despite the mixed bag discussed above and the failure or near failure in a couple of cases, I think that considerable progress was made in the Sloan program towards the goal of showing that it is possible to create such a two-year program - indeed, many such - which will retain the important intellectual values of the traditional two-year program and which, at the same time, will not put students who will need calculus in later courses at a significant disadvantage.

- it will be a long time before there is an agreed upon syllabus for a two-year sequence which includes both discrete mathematics and calculus; this is as it should be because we need to let many flowers bloom in order to determine which can develop into hardy perennials.

- among the key questions which still need answering are the roles of linear algebra and differential equations in the curriculum of the first two years; in particular:
  ◦ Even if you agree that linear algebra naturally falls under the discrete rubric, should the linear algebra in the first two years take on a more computational, less vector space flavor than has been traditional? How does the traditional abstract, vector

space approach square with the trend to less abstract, more "computational" calculus courses?

○ Given that some exposure to differential equations in the first two years is appropriate, how much should there be, how computational should it be and is the inclusion of this material in the same package with some linear algebra sensible?

- a key question is which topics in the traditional calculus sequence would as well or better be taught in a course on discrete mathematics; sequences and series seem obvious candidates but there may be others; a related question is what topics in the calculus sequence might be excised altogether or postponed until later courses, a question I come back to in the next section.

Symbolic Mathematics Systems

Only two of the six institutions - Colby and St. Olaf - experimented with the use of symbolic mathematical systems (sometimes called computer algebra systems [2]) in their calculus courses. One reason for doing this which is emphasized in both reports is "as a tool for helping students realize a deeper conceptual understanding" (to quote from the Colby report) of the mathematics they are studying. This certainly is an important reason for using such systems and I refer you to the reports of these two colleges for more information on this. Little stress is laid in these reports on the impact of such systems on the syllabus itself, a matter I will consider here.

It is ironic that the trend to emphasize a more mechanical, less abstract approach to calculus has occurred at just the time when mechanical devices were being developed to do the symbolic manipulations automatically. In the first two years of college mathematics, as in secondary and elementary mathematics, we are going to have to come to grips with the fact that there will soon be little or no value in and of itself of being a good manipulator of symbols. Within a very few years, ten at most, there will be inexpensive hand-held devices which can differentiate any combination of elementary functions and which will be able to integrate symbolically as well as you would like any sophomore to be able to do. The recently introduced HP-28C and HP-28S are just the first of a series of calculators which will become progressively more powerful and cheaper. This does not mean that students will no longer need to develop "symbol sense" any more than arithmetic calculators mean they do not need "number sense". (In fact symbol sense like number sense may be more important than it has been because of the availability of

mechanical devices to do the manipulations.) But it does mean that students no longer have to be taught to be skilled manipulators of the symbols themselves.

This creates an opportunity and a challenge for all teachers of calculus and discrete mathematics. The opportunity is to delete or decrease the emphasis on topics which have little value beyond their usefulness in symbol manipulation. The standard example of this is various of the techniques of integration (but not integration by parts). A related opportunity is to reduce the time students spend on drill and practice to develop symbol manipulating skills which can be done better by machine. The challenge is to teach students to understand the mathematics you are teaching without requiring them to become able symbol manipulators. Can you learn what a derivative or integral is without becoming very adept at the mechanical manipulations? I think you can, but I have no evidence just as there is little or no evidence that developing good symbol manipulation abilities helps develop understanding. Indeed, it could be argued that the increasing emphasis on mechanical skills in calculus has been at the expense of developing much understanding by students of the underlying concepts.

For much more depth on the teaching of calculus I recommend the proceedings of a symposium on the teaching of calculus held in New Orleans in January 1986 [1].

Discrete Mathematics

In the remainder of this book you will find discussed six different approaches to teaching discrete mathematics in the first two years of college mathematics. There are many points of similarity in the approaches and many differences as well. There is no revealed truth to be found on a syllabus for a one or two semester discrete mathematics course and there will not be for some years. As noted above, it is right for many flowers to bloom. Here is a summary of the points of general agreement among the various schools on topics to be included in a discrete mathematics course:

Algorithms - almost everyone seems to agree that, one way or another, algorithms should play an important role in discrete mathematics course.

Graph Theory - both because its applications are so important and because it gives rise to so many nice algorithms, this is almost universal in a discrete mathematics courses

Combinatorics - at least simple counting problems are also almost universal; for a dissenting point of view see the piece by Steven Bellenot in the Florida State report.

Proof by Induction - this is not mentioned explicitly in some of the reports but I think you can take it as given that you cannot teach a discrete mathematics course without frequent recourse to induction.

Recurrence relations-difference equations - a very common but not quite universal topic which sometimes includes some material on generating functions.

Logic - increasingly some logic, sometimes quite a bit, seems to be a part of discrete mathematics courses; in some cases this takes the form of no more than an introduction to Boolean algebra.

Introductory set theory - many textbook authors use this as an introductory topic intended mainly for review and as an aid to easing students into discrete mathematics.

Beyond this there is a fair amount of idiosyncrasy with such topics as discrete (i.e. computer) number systems, discrete probability, fuzzy sets, modular arithmetic and the summation calculus finding their way into a course or, at least, a textbook.

Which brings me naturally to the topic of textbooks. Until about 1983, there just were no textbooks appropriate for a discrete mathematics course at the freshman or sophomore level. Since then, however, at least 30 books intended for this market have been published and plenty more are on the way. They tend to be quite different in approach, level or topics covered or all three. Four of the six schools whose experiences are reported in this book have faculty members who have published discrete mathematics texts. For a list of recent texts see the end of the Report of the MAA Committee on Discrete Mathematics in the Appendix.

Even if you have found a satisfactory text for a discrete mathematics course, you may not feel comfortable assigning problems or making up tests if you have not taught discrete mathematics before. One of the things which readers of this book should find particularly valuable is the problem sets and examinations which are a part of most of the reports in this book. Both the problems for homework and those on tests encompass a wide variety of topic areas and a considerable range of difficulty. Of particular interest should be the analysis of student performance in the Florida State report.

Conclusion

It is still too soon to say what will be the eventual place of discrete mathematics in the "standard" American mathematics curriculum. All that seems clear now is that the next decade will see a substantial increase in the amount of discrete mathematics taught in the first two years of college mathematics. We may even get to see some experiments in integrated calculus - discrete mathematics courses, surely the most intellectually satisfying, if also the pragmatically most difficult, approach. In any case, I think it behooves all college teachers of mathematics to think about the place of discrete mathematics in the curriculum as a worthy topic on its own and in relation to calculus. The reports in this volume should help in that assessment.

References

1. Douglas, Ronald (editor) (1986): Toward a Lean and Lively Calculus, MAA Notes #6, The Mathematical Association of America.

2. Hosack, J.M. (1986): A Guide to Computer Algebra Systems, College Mathematics Journal, Vol. 17, pp. 434-441.

3. Ralston, Anthony (1981): Computer Science, Mathematics and the Undergraduate Curricula in Both, Amer. Math. Monthly, Vol. 88, pp. 472-485.

4. Ralston, Anthony and Gail S. Young (editors) (1983): The Future Of College Mathematics: Proceedings of a Conference/Workshop on the Future of the First Two Years of College Mathematics, Springer-Verlag, New York.

Colby College

Prepared by Don Small

The goal of the project at Colby was to develop a freshmen-sophomore curriculum that would provide equal weight to both discrete and continuous mathematics. The rationale for adopting this goal is the recognition of the interplay between discrete and continuous mathematics that underlies the development and application of most of mathematics. A natural and basic problem solving approach is to form a sequence of approximations which (hopefully) converges to a solution. Consider preparing for a piano recital or developing the expression for the area of a circle. Although each approximation may belong to the area of continuous mathematics, the sequence belongs to discrete mathematics. The mathematical models in the natural sciences are primarily continuous while those in the social sciences are primarily discrete. In both situations, however, the major computational tool is the digital (discrete) computer. We feel that the core program should recognize these two complementary branches of mathematics and their interactions.

Our approach was to develop a two-semester calculus sequence and a two-semester discrete sequence. The two sequences were to be independent of each other enabling students to take them in either order. We viewed the task of teaching calculus in one year as the crucial hurdle in bringing about a wide agreement among college teachers to include discrete mathematics in the freshmen-sophomore core on an equal footing with calculus. For this reason, along with evidence that several individuals or groups were developing discrete mathematics courses and texts, we focused our primary attention on developing a two-semester calculus course.

Two-Semester Calculus Sequence

Colby College has need for two separate calculus programs, one for students who have studied calculus in high school and one for those who have not. At present, the numbers in these two groups are approximately equal. However, the first group is steadily increasing in size and will clearly become the dominant group in terms of numbers of students. Nationally the number of students studying calculus in high school has increased at a rate exceeding 10% during the last decade. Furthermore, over a third of the freshmen entering a calculus course in college in 1985 had already received a grade of B- or higher in a high school calculus course.

The intended audience for our new calculus course were those students who had already studied calculus in high school or who would be classified as "honor students."

The challenge of fitting all of the essential topics in a standard three-semester calculus course into two semesters was met by:

1. Integrating the single and multivariable calculus in a unified treatment (rather than the standard approach of completing single variable before taking up multivariable calculus).
2. Emphasizing conceptual understanding and development.
3. Using the "Approximation Process" as a simplifying and unifying theme throughout the course.
4. Using a sequential approach to limits.

Starting in the summer of 1983, a new text, Calculus of One and Several Variables: An Integrated Approach was developed. A course sequence based on successive versions of this text has been offered at Colby College since the fall of 1983. The heavy emphasis on sequences throughout the course, particularly sequences of approximations, lends a "discrete flavor" to the course. The interface between discrete and continuous mathematics is best seen in the limit of a sequence. This course satisfies the four conditions recommended by the Calculus Articulation Panel of the Mathematical Association of America for calculus courses for students who have studied calculus in high school and have not received advanced placement. These conditions are:

1. Acknowledge and build on the high school (calculus) experiences of the students.
2. Provide necessary review opportunities to ensure an acceptable level of understanding of Calculus I topics.
3. Provide a course that is clearly different from high school courses. (So that students do not feel that they are essentially just repeating their high school (calculus) course.)
4. Provide an equivalent of a one course advanced placement.

Utilizing a minimum level of "mechanical understanding" from high school calculus allowed for more time (in comparison to standard courses) for motivation and development of concepts. Item two above was easily satisfied by the fact the course covered all the essential topics in a standard three-semester course. The third item is the most crucial one for students who have studied calculus in high school. The integration of several variable calculus throughout the entire course presented the students with new material beginning with the first class. Furthermore the emphasis on conceptual understanding provided the course with a different "flavor". Finally, the fact that the course was two semesters automatically provided the students with a one-semester advanced placement.

The results of the first year were not spectacular. The "finishing rate" (percentage of students beginning the first semester that finished the second semester) was comparable to that in the regular calculus sequence. However, the percentage of students selecting to major in mathematics was much higher in the new course. An "open admissions" policy into the new course and a text that was too hastily written were major factors affecting the results during the first year. The results of the second year were much more impressive, maybe even spectacular, when compared to similar classes of students in previous years. For five years, 1978 to 1982, Colby College offered an experimental section of Calculus II during the fall semester for students who had studied calculus in high school at least through techniques of integration. The first half of the course was devoted to a review of concepts covered in Calculus I. The level of rigor and many of the course goals (learning how to learn, understanding of concepts, appreciation of mathematics) were the same in both the new and the previous experimental course. The selection process was also the same, namely student option. The major differences in the two courses were the integration of single and multivariable material and a text that was written to be compatible with the goals of the new course. To a visitor, the greatest difference would have been the positive, inquisitive, and "willing to work" attitude that characterized the students in the new course in stark contrast to a negative and reluctant attitude of many of those in the previous experimental course.

The previous experimental program was stopped after 1982 for two reasons: first, the ruinous attitude of "I already know the calculus" which pervaded the class and made learning very difficult, and second, the results in the experimental section were not much better than those in the regular calculus course. The following table lists the results of the average of the four years of the experimental course and the 1984-85 new course all of which were taught by the same instructor.

	"Experimental"	"New"
No. of students (September)	20	41
% of students who took a second mathematics course	47	60
% of students who took a third mathematics course	27	37
% of students who chose to major in mathematics	6	30
% of students who chose a combined major of mathematics and another discipline	17	7
% of students who chose a science major	10	12
Attitude	Negative	Positive

As one would expect, the comparison is even more striking between the results of the new course and those in the regular calculus course. For example, only 1% of the students in the regular course chose to major in mathematics.

The number of students taking the new course in the fall of 1985 increased by approximately 50% over the preceding year. In the fall of 1986, the number of sections for the new course increased from two to three. This course had become (within three years) the standard calculus program for students majoring in mathematics or science. An interesting and very satisfying side effect has been the marked improvement in the standard calculus program (measured in terms of attrition rate) as a result of having a more homogeneous class of students.

The Appendix contains the table of contents for the new calculus course and also a sampling of examinations and quizzes.

Computer Algebra Systems

The MAPLE computer algebra system was used on an experimental basis in the new calculus course in 1984-5. The results were very encouraging. This technology is viewed as a tool for helping students realize a deeper conceptual understanding rather than as a time-saving device. Concepts can be discussed differently when there exists a "genie" for the computations in the form of a computer algebra system. Approximations and error bound analysis are of fundamental importance in mathematics. However, this analysis is only lightly touched upon in elementary calculus because of the extensive computations that are often involved. How many times do we ask students to approximate an integral and determine an error bound using Simpson's Rule? The answer, "few." The reason is the difficulty (tediousness) of computing the fourth derivative. Thus, in this situation, the algebra is the restricting influence on the both the type and level of inquiry. In a more general sense, algebraic restrictions are a prime cause for the dependence of our elementary courses on closed form solutions. With access to computer algebra systems, the emphasis can be shifted from closed form to open-ended problems. In particular, approximation and error bound analysis can become a central issue in elementary calculus. Other examples where the use of a computer algebra system can result in a significant change include: graphing as a means of "directing" the analysis rather than just the converse, approximating zeros, convergence of power series, establishing numerical integration as the "norm" with closed form integration considered as a special case.

One of the most important contributions that a computer algebra system can contribute to learning is the influence that it has on the attitude of students with regard to the role of computation in mathematics. Time spent on an activity is often regarded as an indication of the importance of that activity. Since students spend most of their "mathematics time" carrying out routine computations, their tendency is to view mathematics as a "bunch of formulas" (to be memorized) and "to do mathematics" is to compute. ("I can do the mathematics; it's just the definition that I don't understand.") Relegating routine computation to a computer can "free" the student to experiment and conjecture, to "what if," and to "check out" solutions. This type of activity will downgrade the importance of computation in the students' minds and help convince them that the appropriate focus should be on concepts and processes rather than on routine algorithmic manipulations.

Two Semester Discrete Sequence

The difference in mathematical maturity between students who had completed the new calculus program and entering freshmen who do not take calculus dictated the need for two discrete sequences, one at the freshmen level and one at the sophomore level. The sophomore sequence consists of a semester of linear algebra ("Introduction to Linear Algebra" by Bernard Kolman, third edition) and a semester of combinatorics ("Applied Combinatorics" by Alan Tucker, second edition). The existing standard linear algebra course has been replaced by a new course that emphasizes the importance and the wide applicability of linear models. This course reflects the influence of linear numerical analysis on the methods and theory of linear algebra. The prominent role of modern linear algebra in the discrete sequence rests on the importance of the interplay between theory, numerical techniques, analysis of numerical computation, and modeling. Linear models are the primary tools for analyzing the large, complex systems of our society. A large part of the appeal of linear algebra is due to the simplifying and clarifying nature of the theory for these large, complex systems. As Alan Tucker says, "This is what mathematics is really about, making things simple and clear."

At the freshmen level, the goal is to have a two-semester course that emphasizes a modeling approach and integrates linear algebra through both semesters. Students completing this course would have two options for continuing their mathematics study through their sophomore year. Those who wish to major in mathematics or a natural science would be advised to take the new two-semester calculus program. It is expected that the mathematical maturity gained from the freshmen discrete course will compensate for the lack of prior work in calculus when entering the new calculus program as sophomores. Those students planning on majoring in the social sciences or humanities would be advised to take a one-semester calculus course followed by a semester of elementary probability and statistics.

We were surprised at the difficulty of attracting students into a freshmen level discrete course. There seems to be a very strong feeling among our freshmen that, if one takes a mathematics course, it should be a calculus course. In fact, raising the question of "Why take calculus?" appears to be received by freshmen as a strange and senseless question. Clearly an "educational campaign" needs to be waged among students and faculty, particularly freshmen advisors, concerning the role of discrete mathematics in a liberal arts college. This will take time and will require developing a more coherent course than we now offer.

Summary

The entire curriculum project and the personnel involved have received strong positive support from our colleagues within the mathematics department as well as from the College's administration. The College provided a "one course released time" for two semesters to help enable the writing of the new calculus text. We were also given a "free" hand in selecting students and times for our experimental classes.

We are very pleased with the success of our two-semester calculus sequence. This sequence is viewed as the key for revitalizing our mathematics program. The prospects of utilizing computer algebra systems to enhance the effectiveness of instruction in both calculus and linear algebra is very exciting. We believe that these systems provide the opportunity for making major reforms in undergraduate mathematics instruction. We are looking forward to the success of our new unified sophomore-level linear algebra course which was introduced in the fall of 1986. Although our freshmen level discrete mathematics-sophomore linear algebra sequence is not adequate, we remain optimistic that a successful new freshmen sequence can be developed.

In general, we feel that we have succeeded in providing a balance between discrete and continuous mathematics in our core program for the majority of students who take four semesters of mathematics in their freshmen-sophomore years.

In addition to the new calculus and discrete sequences, a new one-semester calculus course has been introduced in our curriculum. The effect of these new courses (along, with the existing standard calculus course and the elementary statistics) is to present a wide selection to students for "mixing and matching" discrete and continuous mathematics courses whether they take two, three, four, or more semesters of mathematics. The number of choices increases the importance on informed advising and thus the need to address the question of the role of discrete mathematics throughout the college curriculum.

The "Sloan curriculum project" continues at Colby College in three areas, even though the funding period has ended.
1. Development of a text by Alan Tucker and Don Small for a unified linear algebra course.
2. Development of a computer algebra system supplement by Don Small and John Hosack.
3. Development of a two-semester freshmen level discrete mathematics course that emphasizes modeling and integrates linear algebra through both semesters.

We wish to express our appreciation to the Alfred P. Sloan Foundation for providing funds to initiate this freshmen-sophomore level curriculum project.

Appendix

Calculus of One and Several Variables:
An Integrated Approach
by Don Small and John Hosack

Table of Contents

Volume 1

This course is taught in two thirteen week semesters with four 50 minute classes per week. There are usually three tests, a final examination, and several 15 minute quizzes given each semester.

The following test materials from 1984-85 include the first test of Semester I, a quiz and first test from Semester II, and the final examination from Semester II.

Math 123 - Test #1

I. True (T) - False (F)
1. $\lim_{n\to\infty} |s_n| = 0$ implies that $\lim_{n\to\infty} s_n = 0$.
2. A bounded monotonic sequence is a convergent sequence.
3. A sequence in R_n converges if and only if it converges component-wise.
4. The set of points $[1,3] \times (2,3]$ is a closed set in R^2.
5. A strictly increasing function has an inverse function.
6. An affine function is a linear function.
7. A polynomial is a linear function.
8. If $f : R \to R^2$, then the graph of f is contained in $R \times R^2$.
9. An unbounded sequence is divergent.
10. Every bounded set contains its least upper bound.

II. Give an example for each of the following conditions.
1. A bounded set that is not closed.
2. A function $f : R^2 \to R^2$ whose domain $\neq$ source.
3. A one-to-one function $g : R \to R^2$.
4. An onto function $h : R^3 \to R$.
5. A monotonic function.

III. Problems
1. The graph of $y = \tan(x)$, $-\pi/2 \times \pi/2$ is:
   Sketch the graphs of the following relations
   a. $y = \tan(x) + 1$
   b. $y = \tan(x + 1)$ $-\pi/2 \le x \le \pi/2$
   c. $y = \tan(2x)$.
2. Sketch the graph of $z = 1 + 2\cos\theta, 0 \le \theta \le 2\pi$.
3. Find a sequence that converges to 2/3.
4. Determine convergence or divergence.
   a. $a_n = (n^2/(n+1), 2, 1/n)$
   b. $b_n = (n^3 + 3n^2 - 7)/(6n^4 + 2n)$
   c. $c_n = \sin(n)/n$

5. Prove, by the definition, that the limit as $x \to -1$ of $f(x) = (x^2 - 1)/(x + 1)$ is -2.

IV. Prove or disprove each of the following statements.
1. $\lim(x^2 - y^2)/(x^2 + y^2) = 1 \quad (x, y) \to (0, 0)$
2. $\lim_{x \to \infty} \sin(x + 2)/(x + \cos(x^2 - 1)) = 0$
3. $f : R \to R^2$ defined by $f(t) = (t^3, t^2)$ is a one-to-one function.

Math 124 - Quiz #2

I (10) 1. (True-False) Every polynomial is an integrable function.
2. (True-False) If f is a positive and piecewise continuous function over [2,5], then $\int_5^2 f(x)dx < 0$.
3. (True-False) If $\int_a^b f(x)\,dx \geq 0$, then $f(x) > 0$ for all x in $[a, b]$.
4. (Cross out all unreasonable answers). $\int_1^2 (x^2 + 2)dx =$
   a. 1.9
   b. 0
   c. 3.7
   d. 5.2

5. (Cross out all unreasonable answers)
$\int_0^3 \sqrt{(1 + \cos(x))}\,dx =$
   a. 2
   b. 8
   c. 0
   d. -2

II. (6) Using the Trapezoidal Rule with its error bound, approximate $\int_0^\pi \sin(x)\,dx$ and compute an error bound. Use a regular partition with $n = 4$.

III. (4) Find the average value of f over $[-2, 4]$ where

$$f(x) = \begin{cases} 2, & \text{for } -2 \leq x < 0 \\ -1, & \text{for } \quad 0 \leq x < 1 \\ 5, & \text{for } \quad 1 \leq x < 3 \\ -2, & \text{for } \quad 3 \leq x \leq 4 \end{cases}$$

Math 124 - Test #1

I. (20) True - False
1. A C^1 function is a function that has a continuous first derivative.
2. Every bounded function over a bounded interval is integrable.
3. If $\int_a^b f(x)\,dx \geq \int_a^b g(x)\,dx$, then $f(x) \geq g(x)$ for all x in $[a, b]$.
4. $\int_a^b \sum_{i=1 \text{ to } n} c_i f_i(x) = \sum_{i=1 \text{ to } n} c_i \int_a^b f_i(x)\,dx$.
5. $\int_2^2 f(x)\,dx = 0$.

II. (12) Give an example or show why no example exists.
1. A discontinuous function defined over $[0, 4]$ that is integrable.
2. An integrable function f defined over $[0, 3]$ such that $|\int_0^2 f(x)\,dx| < \int_0^3 |f(x)|\,dx$
3. An integrable function f defined over $[0, 2]$ whose average value over $[0, 2]$ is not equal to $f(c)$ for any c in $[0, 2]$.

III. (9) Cross out the unreasonable answers.
1. $\int_0^2 \sin(x) \cos(x)\,dx$
   a. 3
   b. -2
   c. π

2. $\int_0^2 \sin(x)\cos(x)\,dx$
   a. -2
   b. 0
   c. 2
3. $\int_0^2 \cos(2x^3 + 3)\,dx$
   a. -2
   b. $1/2$
   c. 3

IV. (8) Describe the rectangle having vertices $(2,2)$, $(7,2)$, $(2,5)$, $(7,5)$ by means of a cross product.

V. (8) Define a (parametric) function whose range is the graph of $y = x^3 + 3$, $1 \le x \le 4$.

VI. (20) Let R denote the region bounded by the graphs of $f(x) = x^3$, $g(x) = x^2$, and $x = 3$. Determine a value of n such that the area of R is approximated with accuracy of 0.01 when
   a. The approximation is a "rectanglular" approximation using a regular partition with n subintervals.
   b. The approximation is a "trapezoidal" approximation using a regular partition with n subintervals.

VII. Consider the curve C described by the range of $f : [0,1] \to R^2$ where $f(t) = (t, \sqrt{t})$.
   a. Determine the integral expression for the surface area generated by revolving C about the x axis.
   b. Determine the integral expression for the solid of revolution obtained by revolving the region bounded by the curves: C, the x-axis, and $x = 1$ about the x-axis.
   c. Which is larger, the surface area in part a. or the volume in part b.? Why?

Math 124 - Final Exam

I. True (T) - False (F)
   1. If $\int_a^b f(x)\,dx \ge 0$, then $f(x) \ge 0$ for $a \le x \le b$.
   2. $e^x > x$ for all x.
   3. If f is a continuous function, then $\int_a^x f(t)\,dt$ is a differentiable function.
   4. $[e^{2x}(3x+4)/(x^2 - 7x)] + C = \int [e^{2x}(3x+4)/(x^2 - 7x)][2 + 3/(4x+4) - (2x - 7)/(x^2 - 7x)]\,dx$.
   5. The function $g : R^2 \to R^2$ defined by $g(x,y) = x^2 - 3\log|xy|$ is a potential function for $F : R^2 \to R^2$ defined by $F(x,y) = (2x - 3/x, 2x)$.
   6. If $\lim_{n\to\infty} a_n = 0$ then $\sum_{n=0 \text{ to } \infty} a_n$ converges.
   7. If the sequence of partial sums of a series is bounded, then the series is convergent.
   8. $|\sin(x) - (x - x^3/6)| < 1/100$ for $0 \le x \le 1$.
   9. The Taylor series for $f(x) = e^x$ about $x = 0$ converges for all values of x.
   10. The Taylor series for $g(x) = \log(x)$ about $x = 1$ converges for all values of x.

II. Give an example or show that no example exists.
   1. A function that is not integrable over $[0,1]$.
   2. A function such that $\int_0^2 f(x) = 0$ and f is not the zero function
   3. A function f that is not increasing over $[-2,4]$, yet $g(x) = \int_{-2}^x f(t)\,dt$.
   4. A series whose interval of convergence is (2,4).
   5. A function $F : R^3 \to R^3$ that has $g : R^3 \to R^3$ defined by $g(x,y,z) = x^2yz^2$ as a potential function.

III. Problems.
   1. The graph of a function is shown for $0 \le x \le 10$. For what value of x will $\int^x f(t)\,dt$, $0 \le x \le 10$ attain its maximum value?
   2. Compute the average value of
$$f(x) = \begin{cases} \cos(x) & 0 \le x \le \pi/2 \\ e^{2x} & \pi/2 < x \le 4 \end{cases}$$
   3. Find the global maximum value of $f(x) = \int_0^x (t^3 - t)\sin(t^{10})\,dt$, $-1 \le x \le 1$.

4. Integrate
   a. $\int \sqrt{(2x - x^2)}\, dx$
   b. $\int x^2 \log(x)\, dx$
   c. $\int (x^4 + 4x^2 + x + 2)/(x^3 + 2x)\, dx$
5. Evaluate $\int F$ where $F : R^2 \to R^2$ is defined by $F(x, y) = (2, \log(y))$ and $0 : [0, 1] \to R^2$ is defined by $C(t) = (t, e^t)$.
6. a. Set up the integral expression for the integral of $f(x, y) = \sin(xy)$ over the region bounded by the graphs of $y = 2 - x^2$ and $y = x$.
   b. Using cylindrical coordinates, set up the integral expression for the volume of the solid under the surface $z = (x^2 + y^2)^{3/2}$ and over the region in the plane that lies inside both curves: $r = 1 + \cos(\theta)$ and $r = 1$.
7. Find the interval of convergence of
   a. $\sum_{k=0 \text{ to } \infty} (x - 2)^k / 2^k$
   b. Approximate $f(x) = x^x$ with a second degree polynomial in powers of $x - 1$.
8. Recall that:

$$\int_0^1 [1/(1 + x^2)]\, dx = \tan^{-1}(x)\, |_0^1$$

$$= \tan^{-1}(1) - \tan^{-1}(0)$$

$$= \pi/4 - 0 = \pi/4$$

   a. Find a series that converges to π.
   b. Determine a partial sum of your series in part (a) that approximates with $ACC = 1/10$.
9. Prove the Taylor series for $f(x, y) = x^2 + 4xy - y^2 + 7$ about $x = 1$ and $y = 0$.

IV. Prove or disprove each of the following statements.
   1. If f is continuous and $f(x) > 0$, then $g(x) = \int_2^x f(t)\, dt$ is an increasing function.
   2. $\int_0^1 1/\sqrt{(1 - x)}\, dx$ converges.
   3. $7/12 < \log(2) < 5/6$. (Do not use calculators on this question.)

University of Delaware

Prepared by Ron Baker

Even in the mathematics community we hear the question—"What is Discrete Mathematics?". There is no easy answer. Some have said that it is the complement of "Continuous Mathematics". We all know that calculus is an example of continuous mathematics. We also know that calculus is the basic language of physics, and that physics has been the basis for engineering's wonderful contributions to our lifestyle. As a result calculus has been the cornerstone of the freshman-sophomore university math curriculum. However, the rise of computing (and computers) is bringing a renewed emphasis on the importance of discrete mathematics. Computer Science has had courses in "Discrete Structures for Computer Science" as a response. Business and the Social Sciences have had courses in "Finite Mathematics". Unfortunately these new courses in discrete math have generally been added to the existing calculus courses, with little or no effort to try to create a coherent freshman-sophomore program.

The Alfred P. Sloan Foundation saw the need for the development of model programs which would have a balance between discrete and continuous mathematics. One of the six pilot programs they funded is the Delaware Revision in Undergraduate Mathematics (DRUM) program. The DRUM Program was established on the following assumptions:

(1) The cornerstone role of calculus was not an accident of history, but rather occurred for sound intellectual reasons. Calculus has not overstayed its welcome in the freshman-sophomore curriculum; indeed it should remain in those programs now requiring it. However, our instruction in calculus has probably become stale and inefficient so that revision could be profitable.

(2) Discrete mathematics provides many useful methods for solving practical problems. It also includes the concepts necessary to understand many academic disciplines. Our courses should blend training in the methods with a clear presentation of the concepts. Solving problems is often the first step in being able to articulate the underlying concepts and justify the methods employed. Stating concepts and justifying methods without being able to apply those methods to solve practical problems is a hollow thing. On the other hand, rote application of methods without understanding is a task for machines, not people. We should try to achieve a balance between the two.

(3) Discrete mathematics and calculus are not separate, competing disciplines. Both are mathematics, and the heart and soul of mathematics is its unity and coherence. Geometry seems to be both part of discrete math and of continuous math, and hence the unity of our courses might be realized by giving each a sound geometric flavor.

To understand the objectives of the DRUM Program, and to evaluate its success or failure, one has to examine the courses created by the program. The four courses created by DRUM are:

M210. Introductory Discrete Mathematics.

(3 credits)

Sets, counting, elementary probability, the integers (well-ordering, Euclidean algorithm and Bezout's identity), vector geometry in the plane and space, matrices, conics and quadrics, Markov chains, numerical sequences and series, induction.

M211. Introductory Calculus.

(4 credits)

Limits and continuity; derivatives and techniques for differentiation; mean value theorem, curve sketching, extrema; integrals, antiderivatives, and the fundamental theorem; logarithmic and exponential functions; basic techniques of integration; applications of the integral; inverse trigonometric and hyperbolic functions.

M212. Topics in the Calculus.

(4 credits)

Methods of integration, improper integrals, l'Hopital's rule, power series and Taylor's theorem; parametric equations, polar coordinates, vector-valued functions; basic calculus of functions of several variables; line integrals. Prereq. M210 and M211.

M215. Analysis of Algorithms.

(3 credits)

Introduction to the questions of termination, correctness, and complexity of algorithms; recurrence relations (difference equations) and generating functions; introduction to graph theory and its algorithms (shortest path, minimal cost spanning tree); algorithms for polynomials including Horner's method; numerical algorithms. Prereq. M210 and M211.

The DRUM Program curricula was offered to a sample population of approximately 250 students in 1984–85. The students came from three degree programs, Mathematical Sciences (math, statistics, and math education), Computer and Information Sciences, and Electrical Engineering. All three groups are required to take the first discrete math course and both semesters of the calculus. Only the CIS majors were required to take the M215 Analysis of Algorithms course (approximately 10% of the M215 class were Math majors taking it as an elective), although EE is considering requiring this course for a concentration in Signal Processing. Approximately two-thirds of the initial population were freshmen taking both a discrete math course (M210) and a calculus course (M211) in their first semester at the university. About 20% of the initial group received credit for one semester of calculus through advanced placement, and a group of students of comparable size entered the DRUM program courses in their second semester after first taking a precalculus course.

Observe that the first semester of discrete math (M210) not only introduces topics from this area because they have importance in their own right, but also includes topics which are critical to the design of a streamlined two-semester calculus sequence. Thus, this discrete math course includes not only counting and induction, but also vector geometry, conics and sequences/series. The content of M210 is commonly dispersed in the present curricula in Finite Math, the three semester calculus sequence, Discrete Structures for CS, et al. Many students do not study Induction in spite of its fundamental importance (especially for computer science). The content of M211 is similar to a typical first calculus course for Engineering/Physical Science students, but does move into topics which were in the second semester of our old sequence. In M212 the presentation of many topics of the calculus is expedited by prior experience not only with one variable calculus but also with vector geometry, conics and quadrics, and numerical sequences and series. The study of induction and the prior experience with algorithms in M210 is the primary topical preparation for the study of the analysis of algorithms in M215.

The commitment of the Department of Electrical Engineering was so strong for 1984–85 that their faculty insisted that *all* of their students (including advanced placement calculus students) be in the DRUM program

courses, with the typical schedule including both M210 and M211 in the fall of the freshman year and M212 in the spring. The Department of Computer and Information Sciences revised its degree requirements with the following changes: (i) The CIS 240-241 "Discrete Structures for CIS" courses were deleted effective Fall 1984; (ii) The introductory Computer Science sequence (CIS 170-171) is revised (now CIS 180-181) to include material on logic and some graph theory not in M210 (the revised course was first offered in 85-86) and to have high school programming experience (or CIS 105, a Pascal-based course) as a prerequisite; and (iii) Four new courses will be required of CIS majors in place of CIS 240-241 and M241-242 (the first two semesters of the old three-semester calculus sequence). The typical new CIS student took M210 and M211 in the fall, M212 and M215 in the spring of 84-85. With the expected demands of the new CIS180-181 sequence the typical CIS student will now take M210-M211-M215-ST205 as a "sequence" over four semesters in the future (ST205 is a Statistical methods course with M210 as prerequisite). The Department of Mathematical Sciences asked all its majors to participate in the DRUM program, by taking M210 and M211 in the fall (if they have an adequate placement level), M212 and possibly M215 in the spring. Since M215 is not best suited to the needs of all Electrical Engineering students, we plan to develop a two semester sequence in "engineering analysis" to serve as an alternative to selecting M215 and M302 (differential equations) for these students. We had not expected M215 to fit the needs of all EE majors (only more Math or CIS oriented EE majors), but we had overlooked the issue of credit hours. M210, 211-212 are a total of 11 credit hours while the old calculus sequence M241-242-243 is 12 credit hours; hence another course is required to maintain levels prescribed for certification of the program. This is really a great opportunity to introduce topics such as difference equations and complex numbers (which are not in the math courses currently required for EE majors). The Department of Mathematical Sciences has revised an old statistical methods course for a freshman year early experience for students concentrating on statistics. These students presently take this course rather than M215 in the freshman year, and will be joined by sophomore CIS majors in the future.

A preliminary edition of a text for M210 was written and used in 1984-85. Materials for M215 were prepared to supplement the text *Concepts in Discrete Mathematics* by S. Sahni (Camelot Publ. Co., Fridley, MN, 1985). We used A. Simon's *Calculus with Analytic Geometry* (Scott, Foresman and Co., Glenview, IL, 1982) for M211-212.

Summary.

1. The pace of the first semester calculus course M211 was well received by both staff and students even though it covered about 30% more material than the old sequence. However the initial population was largely students who had taken a year of high school calculus, so there may be problems with "average" students in this part of the program.

2. In contrast, the second semester calculus course M212 was somewhat troublesome for both staff and students. The staff expressed concern over the lack of maturity and lack of enthusiasm of the students. Opinion surveys showed that this was a manifestation of several factors—difficulty of M212, overall difficulty of course load (the EE majors had begun Physics), the time the course was offered (early morning class, spring of freshman year). On the positive side the performance of the students was comparable to that typical of the traditional three-semester calculus sequence. Student burnout from having two math courses in the first semester has been offered as a possible explanation for the lack of enthusiasm in M212.

3. The adjustment to the first discrete math course M210 was difficult for first semester freshmen. This was a much harder course than M211; however, at the end of the year it was being rated on a par with M212 for difficulty. We have switched to begin the course with vector geometry, rather than counting and probability, since this material has more of the flavor of high school material. There developed a hard core (possibly 20%) who were upset that they were forced to study discrete math in addition to covering *all* of the calculus. On the positive side opinion surveys also contained statements like, "I personally like the new sequence and think it should continue and possibly replace M241, M242, M243 completely." (M241-2-3 is the traditional calculus sequence).

4. The placement of traditional calculus topics, namely vectors and series/sequences, in the discrete math course was very well received by the students. When questioned about being able to carry over these ideas from M210 to M212 we got responses such as "The chapters on vectors, lines and so forth were very useful and easy to carry over because we went into detail about them" and "M210 was an enormous advantage when series and vectors came up in M212. Almost all topics in M210 were beneficiary (sic) to the course. Instead of vectors (and especially series) being the most difficult topics, they tended to be easy." However, there appears to be a great deal of resistance from the faculty to having part of the calculus carry a discrete course as a prerequisite.

5. When asked about what course was the most appropriate prerequisite for physics, the student responses were spread over the spectrum of M210, M211, and M212. We will be looking at objective data to make sure that our structure does not adversely effect the study of physics.

6. M215 Analysis of Algorithms was easily the most popular course in the DRUM program. This is probably natural since the audience was predominantly CIS majors, and this is "their material". The staff (and students) feel that a much better text could be written for this course.

7. Our program has only completed two years of instruction so that many issues are yet to be fully examined. We are committed to continuing M210 and M215 in some form, but the calculus M211-212 (and its relationship to M210) is still being evaluated. We could revert to our three semester calculus sequence M241-242-243 rather than the new M211-212. It appears at this time that a majority of math faculty favor introductory discrete math courses being independent of introductory calculus (no cross-over prerequisites), and not offering a two semester calculus sequence. One of the primary reasons for the popularity of this position is the concern over placing new freshmen in two math courses, which was deemed necessary for the engineering students who must begin physics early. It is clear that the two-semester sequence M211-212 would work well for many more students than we have ever advanced placed in our old M242 (and hence went through the calculus in two semesters). Rather than speeding up the coverage of topics in the calculus (our M211-212), we are more likely to enrich the old three semester format with the use of computing and work harder at placing the large numbers of students with high school calculus into its second course as their initial enrollment. With such an approach to the calculus, we will probably delete series and conics from the discrete course M210. However, we would still teach vector geometry as a natural entry to matrices, and sequences so that recursive sequences get the emphasis needed for the study of the analysis of algorithms. This topical overlap with traditional calculus maintains the spirit of the unity of mathematics which we feel must be

in the basic curriculum.

**Annotated table of contents
for M210 text (Baker/Ebert):**

Chapter 1. VECTOR GEOMETRY

Coverage of the basic topics of vector geometry. The material on coordinates is familiar from high school, and thus provides an easier transition into the course than more abstract topics. The spirit of discrete math begins with the concept of an equivalence relation used to define vectors and continues with the numerous distance problems of three-dimensional space where we avoid "formulas" in favor of "procedures". Optional material on linear inequalities is provided for those who wish to study linear programming in Chapter 7.

Chapter 2. MATRICES

This chapter develops the arithmetic of matrices and Gaussian elimination to solve systems of equations. Thus, we move from the informal "procedures" of vector geometry to our first formal algorithms, both Gaussian elimination and the algorithm for computing matrix inverses. Much of the theory is left to an optional section.

Chapter 3. COUNTING AND PROBABILITY

An introduction of basic counting principles with application to equiprobable models.

Chapter 4. NUMBER SYSTEMS AND INDUCTION

Mathematical induction and well-ordering are presented and developed as tools for proving algorithms correct. Another algorithm is added to the student's experience. We wait until our second course to discuss Program Correctness formally, and then after a review of induction (which our students need since induction is a very hard topic for freshmen).

Chapter 5. SEQUENCES AND SERIES

The usual treatment of sequences and series in calculus is modified by beginning with recursive sequences, which play an equal role with sequences given by a formula for the n-th term. Sections 5.4–5.6 may be omitted from this course if they are to be covered in a calculus course. Recursive sequences better illustrate than the usual calculus problems the distinction between knowing whether a sequence converges and knowing what its limit must be if it does converge. This is especially critical for students who will write recursive programs which have some stopping criteria!

Chapter 6. APPLICATIONS

Our course currently uses Sections 6.1, 6.2, and 6.4. The material on conics profits from the availability of (rotation) matrices, but could be left for a calculus (or precalculus) course. Markov chains are a natural last topic in M210 since they use vectors, matrices, probability and sequences.

Chapter 7. LINEAR PROGRAMMING

We omit this material in our course, however, it could be used to combine Finite Math and Discrete Math in one course.

Chapter 8. GRAPH THEORY

This is material which we choose to cover in our second course M215 Analysis of Algorithms; hence the emphasis on algorithms.

The first appendix collects the usual high school material on unions, intersections, Venn diagrams, etc. The second appendix lists the properties of equivalence relations, and can be used if a more formal introduction to vectors is desired.

University of Denver

Prepared by Ronald E. Prather

1. DESCRIPTION OF THE PROGRAM

At the University of Denver, a new two-year curriculum has been implemented, combining discrete and continuous mathematics from the very beginning of the undergraduate experience. This has been accomplished by collapsing the traditional first-year calculus by a factor of two-thirds (to make room for an elementary discrete mathematics course), and by distributing the second-year linear algebra over the differential equations course and a second course in discrete mathematics.

A. The New Two-year Curriculum.

The new two-year curriculum is outlined in the following table, with course titles (for academic quarters) and with topic headings of the new discrete mathematics courses indented within the first-quarter titles. [Most importantly, one must understand in (1) of the freshman sequence, that the term "calculus" is being used in the broader sense, as a "method of reckoning or calculating."]

THE FRESHMAN SEQUENCE

(1) Discrete Calculus
   - Intuitive Set Theory
   - Deductive Mathematical Logic
   - Discrete Number Systems
   - The Notion of an Algorithm

(2) Differential Calculus
(3) Integral Calculus

THE SOPHOMORE SEQUENCE

(1) Linear Algebra and Combinatorial Mathematics
   - Polynomial Algebra
   - Computational Linear Algebra
   - Graphs and Combinatorics

(2) Linear Algebra and Differential Equations
(3) Multivariable Calculus

B. Schedule of Courses

The new freshman sequence was offered in both the 1983-84 and 1984-85 academic years. Each time, a conventional one-year calculus sequence was offered in parallel, primarily for engineering and physics students. The new sophomore sequence was offered for the first time in the 1984-85 academic year, open to students who had completed the new freshman sequence and to others as well.

2. GOALS OF THE SLOAN PROPOSAL

Through the introduction of this new curriculum, the University of Denver had suggested that the following goals might be achieved:

A. New Text on Elementary Discrete Mathematics

It was anticipated that the indented topic headings of the discrete mathematics courses outlined above would become the chapter titles of a new text, *Elements of Discrete Mathematics* (Ronald E. Prather), Houghton Mifflin, 1986). Designed for freshman-sophomore students and aimed at an intellectual level equivalent to "the calculus," the book would be intended to have wide appeal and would be unique in its emphasis on the notion of an algorithm as central to the development.

B. Supplementary Curriculum Materials

It was proposed that a set of "algorithmic modules" would be written, designed to enrich the continuous topics of the curriculum and to work toward integrating the two divergent streams of mathematical thought into a new pedagogical whole.

C. Experience for the Design of Future Programs

It was suggested that the manner in which the new program had been implemented would facilitate the study of a number of important questions:

- To what extent can the main topics and techniques of "the calculus" be compressed? To what extent can they be treated from an algorithmic point of view?

- Is elementary discrete mathematics more difficult as a first course than is the introduction to calculus? Is it more difficult for freshmen than for sophomores?

- Does an initial exposure to elementary discrete mathematics assist students in understanding calculus?

It was further suggested that even the most tentative answers to such questions would be most useful in the design of future curricula, for mathematics and computer science students particularly, but for others as well.

D. Installation of Permanent Program

It was anticipated that the grant would have the effect of installing the new program as the regular curriculum at the University of Denver.

3. IMPLEMENTATION OF THE PROGRAM

In the first year of its operation, the first two quarters of the new freshman curriculum were taught by Professor Ronald E. Prather; Professor Herbert J. Greenberg taught the third quarter (integral calculus). These responsibilities were essentially reversed in the second year, enabling Professor Prather to initiate the new sophomore curriculum. A doctoral student and lecturer, Paul Myers, assisted both professors throughout the two-year program. In addition, a separate introductory discrete mathematics course for sophomores was offered in each of the two years, taught by Professor Greenberg and others, using the same set of notes as that written for the first freshman course. As indicated previously, these notes treated 'sets', 'logic', 'number systems' (including "induction") and 'algorithms', at an elementary but challenging level.

In order to cover the traditional differential and integral calculus in two-thirds the usual time, a "short calculus" text was selected: S. Lang, A First Course in Calculus, 4th ed., Addition-Wesley. Chapters 3 through 8 were covered in the differential calculus segment and Chapters 9 through 15 (including "series") were covered in the integral calculus segment.

The new sophomore curriculum began with an introduction of three more chapters of the Prather manuscript, treating 'polynomial algebra', 'linear algebra', and 'graph theory', again from an algorithmic point of view. This was followed by a course combining the more abstract facets of linear algebra with an introduction to differential equations. For this purpose, an existing text: A. Rabenstein, Elementary Differential Equations with Linear Algebra, Academic Press, was selected. Concentrating on Chapters 1,3,5, the course centered around the general solution of linear differential equations in order that substantive connections could be made with the foundation in linear algebra. The last course of the sophomore sequence (multivariable calculus) is essentially outside of the scope of the grant program, inasmuch as its content was unchanged from that ordinarily offered at the University of Denver.

4. FURTHER OBSERVATIONS

It is necessary that we briefly discuss two peripheral aspects of the program and comment on their contribution to the overall study.

A. The Traditional Calculus Sequence

At the end of our first year of operation, the "Sloan group" and the group in the traditional calculus sequence took the same standardized test on differential and integral calculus. Taking the test were 26 students in the first group and 53 in the second.

The mean score of the Sloan group was 27 out of a possible 60 points. This places the group in the 43rd percentile nationally. The mean score of the "traditional group" was 32, placing them in the 70th percentile. Some of this difference may be attributed to the fact that the group in the traditional calculus sequence was allowed to use their score (grade) in place of one of their regular exams. Professor Greenberg attributes most of the difference in scores to the additional time spent on "drill" in the traditional group. However, neither group was coached or given any special preparation for the test. It also seems to be a fact that both groups were evenly matched on the basis of SAT or ACT scores.

B. The Sophomore Introduction to Discrete Mathematics

On several occasions throughout the two-year study, we offered a sophomore level course in elementary discrete mathematics, using the same manuscript as that designed for the first freshman course. Unfortunately, the instructors were not always well chosen, neither from the standpoint of an identification with the goals of the Sloan program nor from an affinity or background in discrete mathematics and its computer science applications, generally. Sometimes the class covered little more than one chapter of the notes, so that comparisons would be difficult.

We could interpret all of this as saying that freshman are more capable of absorbing the modern discrete mathematics than are sophomores. But that would surely be stretching the point, and it is probably best to say that the circumstances did not permit a fair study of the question.

5. LONG TERM EFFECT OF THE PROGRAM

In an attempt to foresee the long term effect of the program, we return to a consideration of the four objectives outlined in Section 2 of this report.

A. New Text on Elementary Discrete Mathematics

The final manuscript, Elements of Discrete Mathematics (Prather) was published in 1986. Final reviews were generally favorable and there is little doubt that the book will be well received by a sizable portion of the mathematical and computer science community.

To some extent, the author faced a conflicting set of objectives in the preparation of the manuscript. On the one hand, it is almost certainly the desire and the hope of the Sloan Foundation that a modern and innovative text might result, one that could "move" the curriculum in new and positive directions. On the other hand, the publishers, being notoriously conservative, have tried to steer toward a more conventional text in terms of length, intellectual level, and scope. Necessarily, some compromise was in order. For example, in insisting on a radical shortening of the manuscript, the editors necessitated the complete elimination of the chapter on computational linear algebra. Hopefully, these and other changes will not have a serious impact on the integrity of the text as it was originally conceived.

B. Supplementary Curriculum Materials

It is definitely the case that the traditional continuous mathematics (the calculus, differential equations, etc.) can be treated from a more algorithmic point of view, thus creating more of a bond with modern discrete mathematics. We, in fact, managed to accomplish such a connection on several occasions throughout the program, e.g., in discussing the chain rule of differentiation, in treating the Newton-Simpson approximation to the integral, and in handling the solution of linear differential equations with constant coefficients. Such a treatment seemed to appeal to the students' desire for a rational and meaningful interpretation of these more difficult concepts, something they could actually compute.

On the other hand, we were not able to generate an organized set of these "algorithmic modules" in the limited time available, and certainly nothing of the quality and thoroughness that we would have liked. (The discussion in (D) below gives some hint as to the difficulties we were facing.) It is hoped that some of the other

Sloan studies may have had better luck on this account.

C. Experience for the Design of Future Programs

Here, it is felt that the program proved to be most successful. Speaking only as one, I would say that our experience has established the following points:

- The traditional calculus can be compressed considerably, without a serious sacrifice in understanding or skill.

- The elementary discrete mathematics (of sets, logic, numbers, and algorithms) is not more difficult as a first freshman course than is an introduction to the calculus.

- An initial exposure to elementary discrete mathematics does assist students in their understanding of calculus, their ability to follow proofs, etc.

- The instructor in such a program must be broadly trained in mathematics and well versed in modern computer science applications.

- The elementary discrete mathematics course should be thoroughly integrated with a comprehensive introductory computer science course.

However, it must be emphasized that our study is subject to several interpretations on a number of accounts. If we have chosen to interpret our results in the best possible light in order to call attention to the advantages of this one specific program, that is understandable, of course. But we suggest that the data should stand on its own, so that others may find a somewhat different but equally valid interpretation. In this way, we might be led collectively to the most suitable of future curriculum designs.

D. Installation of Permanent Program

The University of Denver is facing a series of grave financial problems, to the extent that any further consideration of an innovative curriculum for a specific group of students becomes something of a luxury. As a consequence, the future status of the program described herein is very much in doubt.

APPENDIX I

TABLES OF CONTENTS

ELEMENTS OF DISCRETE MATHEMATICS, Ronald E. Prather, Houghton Mifflin, 1986

Chapter 4-The Notion of an Algorithm
Chapter 5-Polynomial Algebra
Chapter 6-Graphs and Combinatorics

A FIRST COURSE IN CALCULUS, Serge Lang, Addison Wesley, Fourth Edition, 1981

Chapter 1-Numbers and Functions
Chapter 2-Graphs and Curves
Chapter 3-The Derivative
Chapter 4-Sine and Cosine
Chapter 5-The Mean Value Theorem
Chapter 6-Sketching Curves
Chapter 7-Inverse Functions
Chapter 8-Exponents and Logarithms
Chapter 9-Integration
Chapter 10-Properties of the Integral
Chapter 11-Techniques of Integration
Chapter 12-Applications of Integration
Chapter 13-Taylor's Formula
Chapter 14-Series
Chapter 15-Vectors
Chapter 16-Differentiation of Vectors
Chapter 17-Functions of Several Variables
Chapter 18-The Chain Rule and the Gradient

ELEMENTARY DIFFERENTIAL EQUATIONS WITH LINEAR ALGEBRA, Albert L. Rabenstein Third Edition, Academic Press, 1981.

Chapter 1-Introduction to Differential Equations
Chapter 2-Matrices and Determinants
Chapter 3-Vector Spaces and Linear Transformations
Chapter 4-Characteristic Values
Chapter 5-Linear Differential Equations
Chapter 6-Systems of Differential Equations
Chapter 7-Series Solutions
Chapter 8-Numerical Transforms

APPENDIX II

SYLLABI FOR THE NEW CURRICULUM

21-194 DISCRETE CALCULUS. Introduction to discrete mathematics. Intuitive set theory; sets, relations, and functions, elementary counting techniques. Deductive mathematic logic; axiomatic theories, proof techniques, truth tables. Discrete number systems; induction and recursion, integers and rationals, base conversion, computer number systems. The notion of an algorithm; recursive algorithms, correctness and complexity theories. Required of all mathematics and computer science majors. 4 qtr. hrs.

21-195.1,.2 DIFFERENTIAL AND INTEGRAL CALCULUS. Differentiation and integration of functions of one variable, infinite sequences and series; partial differentiation of functions of several variables; illustrations of the algorithmic aspect of the calculus. Required of all mathematics and computer science majors. 4 qtr. hrs.

21-204 COMPUTATIONAL LINEAR ALGEBRA AND COMBINATORIAL MATHEMATICS. Polynomial calculus; algebra of polynomials, division and factor algorithms. Computational linear algebra; vectors and matrices, matrix algebra and determinants, dimension and rank, systems of linear equations. Graphs and combinatorics; paths and circuits, planarity, coloring problems, combinatorial algorithms, introduction to discrete probability. Required of all mathematics and computer science majors. Prerequisite: 21-194. qtr. hrs.

21-207 LINEAR ALGEBRA AND DIFFERENTIAL EQUATIONS. Vector spaces, linear transformations, eigenvalues and eigenvectors. Solution of linear differential equations; special techniques for handling nonlinear problems including numerical methods; mathematical modeling of problems from the physical and biological sciences. Prerequisite: 21-195.2 and 21-204. 4 qtr. hrs.

APPENDIX III

SAMPLE EXAMINATION QUESTIONS

DISCRETE CALCULUS

1. Suppose we are to prove one of DeMorgan's Laws:

$$\neg(A \cup B) = \neg A \cap \neg B$$

We then have two inclusions to verify, namely:

_____ and _____

Fill in the blanks below and provide a brief parenthetical justification for the steps that establish these two inclusions.

Let $x \in \neg(A \cup B)$ Let $x \in \neg A \cap \neg B$

$x \notin A \cup B$ (definition of $\neg$) $x \in \neg A$ and $x \in \neg B$ (definition of $\cap$)

( )	( )
( )	( )
( )	( )
( )	( )

Check the validity of this law in case

$$A = \{1,3,7\} \qquad B = \{2,9\}$$

in the universe $U = \{1,2,3,4,5,6,7,8,9,10\}$, showing each step in your calculations:

$$\neg(A \cup B) = \neg(\{1,3,7\} \cup \{2,9\}) =$$

$$\neg A \cap \neg B = \neg\{1,3,7\} \cap \neg\{2,9\} =$$

2. Let $f : S \to T$ be any function. Define the relation $\approx$ on S by taking:

$$x \approx y \text{ to mean } f(x) = f(y)$$

Show that this is an equivalence relation by establishing the properties:
(E1) reflexive
(E2) symmetric
(E3) transitive

3. Compute the following (show your work):
   (a) number of possible rankings of the "final three" in the Miss America Contest.
   (b) number of D. U. Student identification numbers (letter or digit, followed by four digits).
   (c) number of ways of choosing three months of the year as discount months at a department store.
   (d) number of subsets of a five-element set.
   (e) the first six rows of Pascal's triangle.

4. (A Little Geometry) Let *line*, *point*, and *on* be primitive terms of "geometry" with axioms that include:
   (H1) Any two distinct points lie on one and only one line.
   (H2) Any two lines intersect.

   With the following interpretations:
   point = one of the elements of the set $\{1,2,3,4,5,6,7\}$
   line = any of the following triples of elements:
   $\{1,2,3\}$ $\{1,4,7\}$ $\{1,5,6\}$ $\{2,4,6\}$ $\{2,5,7\}$ $\{3,4,5\}\{3,6,7\}$
   A on l means $A \in l$

   (a) Show that axioms (H1) and (H2) are satisfied. Write "etc." after you have done enough to make your point.
   (b) Draw a "picture" of (the lines of) this geometry:

5. (A Little Algebra) We are going to prove the implication (See instructions below):
   $y^2 > 0, x^2 + 6y^2 - 25 = 0, y^2 + x = 3 \Rightarrow |y| = 2$

   ⓐ _____
   ⓐ _____
   ⓐ _____

$x^2 + 6(3 - x) - 25 = 0$ (substitution)

——————————————   (algebra - couple of steps)

$(x - 7)(x + 1) = 0$ (factor)

——————— v ——————— $(ab = 0 \Rightarrow a = 0 \text{ or } b = 0)$

——————— $\to y^2 \le 0$

————————

————————   (substitution)

$x \ne 7$ [   ]

$x = -1$

$y^2 = 4$ [   ]   These are derived rules

          (substitution)

ⓑ ——————— v ———————   (take square roots)

——————— $\to$ ———————   (definition of absolute value)

——————— $\to$ ———————   (definition of absolute value)

ⓐ ——————————————

    ⓐ   Fill in the assumptions and the conclusions.
    ⓑ   Backward reason.
    ⓒ   Fill in the other blanks at the left and justifications [   ] at the right.

6. Justify the steps (those that need justification) with deduction rules

$$p \wedge q \to (r \to \neg s), r \wedge s \Rightarrow p \to \neg q \vee \neg r$$

$$p \wedge q \to (r \to \neg s)$$

$$r \wedge s$$

$$p \to \neg q \vee \neg r$$

$$p$$

$$q \to s \wedge \neg s$$

$$q$$

$$p \wedge q$$

$$r \to \neg s$$

$$r$$

$$\neg s$$

$$s$$

$$s \wedge \neg s$$

$$\neg q$$

$$\neg q \vee \neg r$$

7. State the law or rule that distinguishes classical logic from constructive logic.

8. Compute the truth table of the following logical expression:

$$((p \vee r) \wedge \neg q) \to (r \wedge \neg(p \wedge q))$$

9. (a) Convert the number 109.8125 from decimal to binary.
   (b) Convert the number 1011011.1101 from binary to decimal.

10. For the following sets:

		finite	countable	uncountable
(a)	$N \times N$			
(b)	all squares of integers			
(c)	Q			
(d)	all well-formed parentheses strings			

fill in the table with "checks" (   ) to indicate whether the set is finite, infinite but countable – having the cardinality of N, or uncountable.

Verify your answer for either (a) or (b)

11. Consider the three functions described below.

$$f : S \to T \qquad\qquad g : Z \to N \qquad\qquad h : N \to N$$

$$g(y) = \begin{cases} 2y & \text{if } y > 0 \\ 2|y| + 1 & \text{if } y \le 0 \end{cases} \qquad h(x) = \begin{cases} \text{smallest} \\ \text{perfect square} \\ \text{greater than} \\ \text{or equal to x} \end{cases}$$

Compute each of the following:

$f(1) =$ $\qquad$ $g(0) =$ $\qquad$ $h(8) =$
$f(2) =$ $\qquad$ $g(1) =$ $\qquad$ $h(9) =$
$f(4) =$ $\qquad$ $g(-1) =$ $\qquad$ $h(10) =$
$\qquad h \circ g(0) =$ $\qquad\qquad g^{-1}(4) =$
$\qquad h \circ g(-100) =$ $\qquad\quad g^{-1}(5) =$

NOTE: g^{-1} is the inverse of g

Give yes or no answers for each of the following:

$$f \qquad\qquad g \qquad\qquad h$$

is one-to-one?
is onto?

12. (a) Provide a proof of the following:

$$p \wedge q \Rightarrow (p \wedge \neg q)$$

(b) You have thus obtained a new "derived rule," call it

[∧D] ————————————————⟶ Fill in the blanks

(c) Use it to give a short proof of the following:

$$(r \wedge s) \wedge t \Rightarrow \neg t \vee \neg (r \wedge \neg s)$$

13. You are going to prove that the proposition:

$$P(n) : \sum_{j=1}^{n} j^2 = n(n+1)(2n+1)/6$$

is valid for all $n \geq 1$.
   (a) Show that P(1) holds.
   (b) What is your "inductive hypothesis" (circle one)

                            for $n = 1$
                            for all $n \geq 1$
        that $P(n)$ holds: for some $n \geq 1$
                            for sufficiently large n

(c) Show that $P(n) \to P(n+1)$ and identify the point at which the inductive hypothesis is used, in showing that

$$\sum_{j=1}^{n+1} j^2 = \frac{(n+1)((n+1)+1)(2(n+1)+1)}{6}$$

14. Rewrite the bisective search

    algorithm bisection $(X, x :$ found, location$)$
    set top and bot to first and last respectively
    initialize found as 'false'
    while top $\geq$ bot and $\neg$ found
        compute mid
        case comparison
            $x < X_{\text{mid}}$: change bot to mid-1
            $x > X_{\text{mid}}$: change top to mid$+1$
            $x = X_{\text{mid}}$: set found as 'true'
    assign location as mid
    as a recursive function (fill in the blanks):
        function search (top, bot)
        if top $>$ bot
            return [   ]
        else
            compute mid
            case comparison
                $x < X_{\text{mid}}$: [   ]
                $x > X_{\text{mid}}$:
                $x = X_{\text{mid}}$:

    returning the index of the location in a list X where x is found (assume it's there).

    Show that the recursive algorithm (function) satisfies the relation:

$$T(n) = 1 + T(n/2) \qquad (n = size(X))$$

HINT: Take a "worst case." view of the if-else and case constructs.

It follows that the recursive algorithm (function) has the order of complexity: (circle one)

$$\log n \quad n \quad n^2 \quad 2^n \quad n/2$$

15. Euler's function (the Greek letter "phi") $\phi(n)$ counts the number of positive integers less than or equal to n that are "relatively prime" to n. (Recall that two natural numbers are relatively prime if they have no common divisor, other than one.) Write a pseudo language algorithm

$$\text{euler } (n : \text{phi})$$

that will compute this number when n is given as input. NOTE: You may call the gcd function in developing your solution.

16. The conventional exchange sort routine is as follows:

    algorithm exchangesort (A)
    for i running from 1 to n i
        for j running from l to $n - i$
          if $A_j > A_{j+1}$
            interchange A_j and A_{j+1}

We can improve its efficiency by using a Boolean variable to detect the fact that the array may be prematurely sorted, i.e., it may be that there were no interchanges necessary on a given pass. This fact can be detected by using a counter – simply counting the number of interchanges on a pass. Show how this technique is implemented (fill in the blanks) in the following:

    algorithm improvedexchangesort (A)
    set unsorted to 'true'
    initialize i at 1
    while _____
        initialize COUNT at 0
        for j running from 1 to $n - i$
          if $Aj > A_{j+1}$

            _____
            _____
      if _____
        _____
        _____

17. At the right, trace the state of the computation (with input $n \geq 1$) for the algorithm at the left. Here, we take $n = 5$

    algorithm what $(n : f)$
    set f to 1
    set m to n
    while $m > 1$
        multiply f by m
        decrease m by 1

n	m	f
5		

Use your experience with the above table to help you "discover" an invariant assertion for the while loop, an assertion of the form:

$$f \cdot g(m) = g(n)$$

for some integer function g.

What form does this assertion take when we exit the while loop?

What is accomplished by the algorithm?

COMPUTATIONAL LINEAR ALGEBRA AND COMBINATORICS

1. A plane passes through the points $(1, 1, 1)$ $(7, -1, 0)$, and $(-8, 5, 2)$. Find a unit normal vector to the plane and the points of intersection of the plane with each of the three coordinate axes.

2. Find a basis for the row space and the rank of the following matrix.

$$A = \begin{bmatrix} 1 & 3 & 1 \\ 2 & 1 & -3 \\ 0 & 2 & 2 \\ 3 & 8 & 2 \end{bmatrix}$$

3. Fill in the table with appropriate checks (   ) for the following sets of vectors.

	space		vectors	
$Q^{2,2}$	$\begin{bmatrix} 0 & 0 \end{bmatrix}$	$\begin{bmatrix} 1 & 0 \end{bmatrix}$	$\begin{bmatrix} 0 & 0 \end{bmatrix}$	$\begin{bmatrix} 0 & 0 \end{bmatrix}$
$Z_3[x]$	$x, x^2, x^3 - 2x, x^3$			
Q^3	$(1, 0, 0), (0, 1, 0), (0, 0, 2)$			
B^4	$(1, 0, 0, 0), (0, 1, 0, 1), (1, 0, 1, 0), (0, 1, 1, 1)$			
Q^4	$(1, 1, 0, 0), (0, 2, 2, 0), (0, 0, 3, 3)$			

	dependent	independent	basis
$Q^{2,2}$			
$Z_3[x]$			
Q^3			
B^4			
Q^4			

4. Use the definition of linear dependence or independence (whichever is appropriate) to verify your check in the last row of problem 3.

$$A = \begin{bmatrix} 1 & 2 & 3 & 4 \\ 0 & 1 & 2 & 3 \\ 1 & -2 & -6 & -7 \\ 4 & 3 & 1 & 2 \end{bmatrix}$$

5. Using the cofactor expansion down the first column, compute the determinant $|A|$.

6. Using Gaussian elimination with "half-pivoting," reduce A to echelon form.

7. Using back-substitution with the result of problem 6, describe (in column matrix form) all solutions to the homogeneous system $Ax = 0$.

$$B = \begin{bmatrix} 2 & 3 & 4 \\ 1 & 2 & 3 \\ -2 & -6 & -7 \end{bmatrix} \qquad C = \begin{bmatrix} 1 & 2 & 3 \\ -2 & -6 & -7 \\ 3 & 1 & 2 \end{bmatrix}$$

8. Compute each of the following:
   (a) $B + C$
   (b) BC

9. The matrix B has the cofactors (signed minors):

   $C_{11} =$ $C_{12} =$ $C_{13} =$
   $C_{21} =$ $C_{22} =$ $C_{23} =$
   $C_{31} =$ $C_{32} =$ $C_{33} =$
   and the inverse
   $B^{-1} = 1/|B|\ \text{adj}\ B =$

10. Use the result of problem 9 to compute the solution (in column matrix form) to the system of linear equations

$$Bx = y \text{ where } y_1 = y_2 = y_3 = 1.$$

$$a(x) = 4x^3 + 6x^2 - 2x - 3$$

11. (Descartes' Rule of Signs) Describe the signs (+ or −) and indicate the conclusion.

signs of $a(x)$				
signs of $-a(x)$				

    $\Rightarrow$ at most _____ positive root (s)
    $\Rightarrow$ at most _____ negative root (s)

12. (Synthetic Division and Evaluation) Perform the indicated evaluations, describe the bottomline of the synthetic divisions and use all of this information to locate the roots.the roots.

    $a(0) =$ _____

    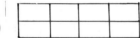

    $-2\rfloor$ ⌊_ $\Rightarrow$ all roots are _____

    $1\rfloor$ ⌊_ $\Rightarrow$ all roots are _____

    (Change of Sign Principle) There is certainly a root where? _____

13. (Rational Roots) List all possibilities for rational roots p/q, cross out those that are disqualified on the grounds of (12), circle one that works, show it with synthetic division.

14. (Quadratic Factor) List all possibilities $b(-2)$, $b(0)$, $b(1)$, circle ones that work and set up the corresponding LaGrange Interpolation Formula for a quadratic factor b.

(−2,)	(0,)	(1,)
(−2,)	(0,)	(1,)
(−2,)	(0,)	(1,)
(−2,)	(0,)	(1,)

$$b(x) = \frac{(\quad)}{(\quad)}\frac{(\quad)}{(\quad)}\frac{(\quad)}{(\quad)} + \frac{(\quad)}{(\quad)}\frac{(\quad)}{(\quad)}\frac{(\quad)}{(\quad)} + \frac{(\quad)}{(\quad)}\frac{(\quad)}{(\quad)}\frac{(\quad)}{(\quad)}$$

$$= \underline{\hspace{4cm}} \quad (\text{ in standard form})$$

15. (Approximate Roots) Complete two lines of the following tabulation, show the synthetic divisions required to evaluate a(mid), and describe the (irrational) root.

left	right	mid	a(mid)
0	1		

root = _____

16. (Factorial Polynomials) Use the Newton Expansion as indicated to convert $a(x)$ to factorial form.

$$\Delta^0 a(x) = 4x^3 + 6x^2 - 2x - 3 \qquad \Delta^0 a(0) =$$
$$\Delta^1 a(x) = \qquad\qquad\qquad\qquad \Delta^1 a(0) =$$
$$\Delta^2 a(x) = \qquad\qquad\qquad\qquad \Delta^2 a(0) =$$
$$\Delta^3 a(x) = \qquad\qquad\qquad\qquad \Delta^3 a(0) -$$
$$4x^3 + 6x^2 - 2x - 3 \rightarrow$$

17. (Use of Fundamental Theorem) Evaluate the following sum (See Problem 16)

$$\sum_{x=0}^{n} (12x^2 + 24x + 8) =$$

18. For the polynomial $a(x) = 4x^3 + 6x^2 - 3$ use synthetic divisions to establish a range within which the roots must lie.

range: _____ < roots < _____

List all possibilities for rational roots p/q, cross out those that are disqualified on the above grounds, circle one that works, show it with synthetic division.

19. Use Gaussian elimination with half-pivoting to reduce

$$A = \begin{bmatrix} 1 & 2 & 3 & 4 \\ 0 & 1 & 2 & 3 \\ 0 & 2 & 3 & 7 \\ 4 & 3 & 1 & 2 \end{bmatrix}$$

to echelon form.

With this result, use back-substitution to describe (in column matrix form) all solutions to the homogeneous system $Ax = 0$.

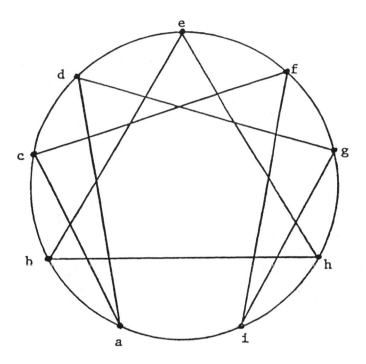

20. Compute the circuit rank.

21. How many edges will there be in a spanning tree?

22. Exhibit a spanning tree.

23. Into how many (r) regions would the plane be divided if the graph could be drawn without any intersecting edges?

24. Show that the graph is nonplanar. (HINT: There are only 5 triangles.)

25. Give the covering number α and a corresponding minimal covering set.

26. Give the independence number β and a corresponding maximal independent set.

The Florida State University

Project Participants

Professor Steven F. Bellenot (Dept. of Mathematics)

Professor John L. Bryant (Dept. of Mathematics)

Associate Professor Bettye Anne Case (Project Coordinator) (Dept. of Mathematics)

Professor Ralph D. McWilliams (Dept. of Mathematics)

Professor Joe L. Mott (Dept. of Mathematics)

Professor Theodore P. Baker (Dept. of Computer Science)

Professor Abraham Kandel (Dept. of Computer Science)

Dr. Joseph F. Hoffmann (Tallahassee Community College)

Contents

PROJECT SUMMARY
B. A. Case

1. Course Background, Parameters and Development.

Since Spring, 1980, computer science majors at The Florida State University have been required to complete a year of discrete mathematics at the freshman-sophomore level not requiring as prerequisite computer programming experience or calculus. Around 1,700 students were taught one of the sequence courses Fall 1982–Spring 1985. Discrete I was added to the mathematics major during the period that the Sloan Discrete Mathematics Project was in proposal form; it has since been added as a requirement in the electrical engineering computer option and for mathematics education. Although a number of mathematics professors would like to encourage other science and engineering areas to require Discrete I, with Discrete I required as a prerequisite to Linear Algebra, interest seems to come primarily as those disciplines need computer science courses for which at least Discrete I is prerequisite. (All computer science courses except first-semester language and service courses require Discrete I.) Success in Discrete I is enhanced by a rather sophisticated prerequisite college algebra including logic, elementary combinatorics and a careful treatment of mathematical induction.

When the course was started there was not a suitable text at the freshman level. The faculty started immediately to develop suitable materials, and the locally written Mott, Kandel, Baker: Discrete Mathematics (Reston- Prentice-Hall) was first used at the beginning of this project, Fall 1983. The revised Second Edition utilizing revision and materials written during the time of this project is now available. (See Second Edition Table of Contents which follows this summary.)

2. Staffing, Content and Course Sequencing.

The number of discrete sections required had already stretched available faculty resources, and a major need at the beginning of this project was the development of materials so that some sections could be assigned to adjunct or teaching assistant instructors, as is the case with calculus. During this project annotated Instructor Notes were developed to facilitate the teaching of this non-traditional course. (These were shared with instructors at many other schools.) Carefully screened teaching assistants and other non-regular faculty instructors have been successful and have greatly enjoyed teaching Discrete I. They report it is challenging to them, and, if they have not otherwise had occasion to do so, often prefer to sit in a section the semester before they teach it. The Instructor Notes are carefully annotated. Each of these instructors has a faculty mentor who has taught the course many times, and the approach to the material as well as tests are discussed carefully.

Some of the faculty other than the authors who taught the course during its first years developed materials and methods which were included in the Instructor Notes. For example, there were at least three variants in the order in which the major topics of Discrete I were taught from the First Edition. The decision of whether combinatorics or graphs will be taught first, after completion of material on logic, functions and proof methods (including induction), is left to the instructor's discretion. Some discussion of the advantages of each is included in the Instructor Notes. The text authors and many teaching the course for the first time teach some combinatorics, then graphs and trees, after the foundations material in Discrete I. (See "Author's Report" by Mott.) This writer teaches just that content, but after experimentation, prefers to teach combinatorics last because the material on graphs and trees seems to be especially appropriate for reenforcement of the foundations material. One of the project participants presents the case for teaching combinatorics last in Discrete I and, in fact, deemphasizing it (See "'Counting' vs 'Proofs'" by Bellenot.)

The inclusion of chapters (topics) from the text in each course is:

Discrete I: 1 (foundations); most of 2 (combinatorics); some of 4 (relations and digraphs); most of 5 (graphs and trees).

Discrete II: 3 (recurrence relations); 7 (network flows); remaining parts of 2, 4 and 5; other topics as time permits.

3. The Writing Project.

In addition to the development of supplemental materials and revisions in ordering and emphasis, there were two major writing projects which are incorporated as new chapters in the Second Edition. (See attached Table of Contents.) The unit on Network Flows (authored by Mott) is a regular part of Discrete II; the material was carefully class tested in note form during the project. Kandel's material in Chapter 8 was developed in a more advanced course and applies fuzzy

sets and techniques to Artificial Intelligence, using discrete mathematics concepts such as graph theory and relations.

4. Discrete I at Tallahassee Community College.

The course has been successfully taught at Tallahassee Community College. Hoffmann reports little difference between his class there and those he has taught at Florida State as an adjunct instructor. The proof-oriented nature of the FSU version of Discrete I is suitable for the TCC population and is, in fact, essential so that those who transfer will not be behind their contemporaries. Hoffmann reports the course to be more challenging and interesting to teach than many. (See attached report by Hoffmann and Case.)

5. Related Projects and Information Dissemination.

Due to Sloan funding, we have been able to send out materials to over 100 mathematicians about our discrete mathematics course with major mailings in August, 1984, and February, 1985. We are in active communication with two-year college faculty having interest in developing discrete courses and would like to hold a month-long summer program for them. Regrettably, the NSF funding structure does not seem to allow this, and other funds do not seem to be available. Case and Mott have each given a number of talks at MAA section meetings and also state-wide meetings attended by high school and community college teachers. Mott teaches discrete mathematics courses to high school students in the summer.

6. Course and Data Evaluation Reports.

Data analysis studies have been carried out by three of the project participants. In an attached summary, McWilliams closely analyzes Discrete II exam item performance. Case takes a broad look at grade distributions and their relationship to student preparation. Bellenot in his report mentioned earlier presents considerable data with test item analysis.

7. Conclusions.

In his "Author's Report," Mott presents opinions, these in the context of his experience in teaching the courses each semester, about developing and reworking material and about mentoring to beginning instructors. He gives specific examples and also describes aspects of the departmental philosophy about the course. The reports attached by Bellenot, Case,

Hoffmann, and McWilliams each include summary information and conclusions. The raw quiz, test and examination materials included in the *Appendix* reflect the emphases of many professors and instructors teaching the two semesters of material. The text revised on the project is flexible and adapts to these concerns and approaches. As project participants we have enjoyed and our students have profited from this close examination of philosophies, methods and results related to teaching discrete mathematics in the first two college years. We hope consideration of our materials will be helpful to others developing these courses.

8. Acknowledgement.

The project participants appreciate the support of the Alfred P. Sloan Foundation; each has enjoyed working on the project.

DISCRETE MATHEMATICS FOR COMPUTER SCIENTISTS AND MATHEMATICIANS
Second Edition

Joe L. Mott   Prentice-Hall
Abraham Kandel   Englewood Cliffs, New Jersey
Theodore P. Baker   1986

Table of Contents

AUTHOR'S REPORT
Joe L. Mott

At the outset, let me say how much I enjoy teaching discrete mathematics; I am sure that most will agree that it is a challenging and rewarding experience for teacher and student alike.

There are singular qualities about the subject matter that set off discrete mathematics from other undergraduate mathematics courses. In a combinatorial problem, for example, only the slightest change in the problem frequently requires a totally new analysis and a completely fresh approach to solve it. Moreover, without requiring previous knowledge of formulas like the volume of a right circular cylinder, the trigonometric double angle formulas, Newton's second law of motion, or the point slope formula for a line, we can pose good problems in discrete mathematics in terms of very common concepts. Discrete mathematics is like number theory (and vastly different from calculus) in that regard. After all, you can hardly discuss anything more basic than properties of integers, or the number of bridge hands of a certain type, or ways to determine the maximal amount of oil that can flow from a set of refineries to a collection of markets along existing pipelines.

Besides that, discrete mathematics is almost totally self-contained. No previous experience is required in trigonometry, geometry, number theory, physics, or any other science. To be sure, exposure to computer science is desirable but not essential. Evidently a solid background in algebra and an open, willing, inquiring mind are the only requisite tools.

Another nice feature about discrete mathematics is its adaptability; it can be adjusted to fit the needs of undergraduate students from all levels regardless of previous mathematical experience. For the last several summers, for example, I have chosen certain topics from discrete mathematics and used them successfully in a science and mathematics camp for bright high school seniors. Also, on occasion, graduate students from other disciplines have enrolled in our discrete mathematics classes.

While it is true that the topics of discrete mathematics are adaptable to students with varying backgrounds, I should hasten to make one equivocation. In my mind there are two levels of discrete mathematics textbooks. All basically contain combinatorics to some extent, graph theory (including modeling of real-life problems), and algorithms. Most contain some discussion of logic (though I believe that the Mott-Kandel-Baker text [12] was the first published to discuss logic and methods of proof). But there is one distinguishing topic that renders certain texts inappropriate for a sophomore level course and that topic is group theory and Polya enumeration theory. In my opinion, any discrete mathematics course that includes much on Polya enumeration automatically becomes a senior level course. Thus, I believe that, if discrete mathematics is offered in any way as an alternative (or as a supplement) to the regular calculus sequence at the sophomore level, then group theory should be omitted as a topic. In that regard, I also believe that Kuratowski's characterization of planar graphs is a beautiful result, but I personally see no use for it in a sophomore course.

After those hedges, let me return to the discussion of the nice features of discrete mathematics. One particularly enjoyable feature is that in a minor (but nevertheless real) way discrete mathematics captures the spirit of mathematical and scientific research. Many times, in order to solve a problem, one must experiment with special cases, look for a pattern, make conjectures, discover counterexamples, refine the conjectures; and so on. G. Polya has discussed in several books this process of scientific reasoning and has illustrated in excellent fashion the spirit of discovery. Indeed, the very interesting book [13] by G. Polya, R. Tarjan, and D. Woods contains several illustrations of how Polya teaches certain topics in combinatorics.

AN EXAMPLE: Allow me here to illustrate one example I always use. I list a few values of the function $f(n) = n^2 - n + 11$.

n	0	1	2	3	4	5	6	7
$f(n)$	11	11	13	17	23	31	41	53

Figure 1.

Then I ask for conjectures about additional values of $f(n)$, and usually the students come up with conjectures like:

(1) $f(n)$ is prime for all integers $n \geq 0$.
(2) $f(n)$ is odd for all integers $n \geq 0$.

After the latter conjecture, I ask for proofs from the class. They usually prove it by cases. Then I offer the observation that since $n(n-1)$ is even, $f(n) = n(n-1) + 11$ will always be odd. But then (2) implies

(3) $f(n+1) - f(n)$ is even for all integers $n \geq 0$. (After probing, the conjecture is refined to $f(n+1) - f(n) = 2n$, a fact the class can immediately prove together.)

After some consideration of (1), someone will see that $f(11)$ is not prime. Then someone may conjecture that $f(n)$ is prime for all integers except $n = 11$. But then we observe that $f(n)$ is divisible by 11 if n is divisible 11. Then someone may conjecture that $f(n)$ is prime for values of n not divisible by 11. But again this is not the whole truth since $f(12)$ is composite. In fact, if $n = 11k + 1$, then $f(n)$ is divisible by 11.

Sometimes a student will conjecture:

(4) The sequence of units digits is the periodic sequence 1, 1, 3, 7, 3, 1, 1, 3, 7, 3, etc.

Of course, this is easily verified by appeal to congruences modulo 10.

After each proof or counterexample, I engage the class to make a new conjecture, to test that conjecture, and to offer a counterexample or give a proof. I treat any conjecture as something worthy of investigation.

After a while, I might lead the class in a new direction. For example, we may note that $f(n)$ is prime for all the values $n = 0, 1, 2, \ldots, 10$ and that $g(n) = n(n-1) + 5$ is such that $g(n)$ is prime for all values of $n = 0, 1, 2, 3, 4$. Then we may conjecture that the function $f_p(n) = n(n-1) + p$, where p is a prime, is such that $f_p(n)$ is a prime for all values of $n = 0, 1, 2, \ldots, p-1$. Then off we go again. (The only primes for which this last conjecture is true are $p = 2, 3, 5, 11, 15, 41$. For more information, the reader may consult [5], p. 38.)

RECOMMENDATIONS FOR NEW TEACHERS.
Frequently, for the person just starting to teach discrete mathematics, I recommend for personal enrichment several well written elementary books including [2], [3], [6], [9], [11], [16], and [17].

Another thing I strongly recommend to new teachers of discrete mathematics is that they choose a textbook with a good solutions manual. Moreover, expect to use the solutions manual more than normal. The first time I taught discrete in the spring of 1980, I had never before used a teacher's manual in any course. In that first course, I used C.L. Liu's great book [10]. Soon I discovered that I could spend a great deal of time on solutions of problems just trying to see whether or not I could expect my class to do them. (There was a solutions manual but I was not aware of that at first.) Many of the problems were, in my opinion, too difficult for sophomores so I started to prepare handouts with additional problems and their solutions. This was the start of the first edition of our text [12]. Incidentally, without the benefit of the Williams College Conference recommendations and based on our intuition as to what was needed and what could be taught effectively, Kandel, Baker, and I came up with a table of contents remarkably in harmony with the suggested contents for a discrete mathematics course.

Let me also suggest that you keep problem banks, one for each section of the book along with completely worked out solutions. Use these for quiz starters. Old examinations don't quite serve this purpose as they usually cover several sections. What I do is this: I keep several worked-out problems on each section different from (but similar to) the problems in the text. I keep these in several large looseleaf notebooks. Then if I want to make an examination covering certain sections and requiring certain ideas, I can go to the section banks, take a problem out of each, check that the solution does in fact require the ideas I want tested, and combine these problems into an examination.

RESOURCE MATERIALS AVAILABLE.
In light of our experience of teaching from Liu [10] and other books, and as part of what we have done with the Sloan grant, we have prepared a packet of problem sets, solutions, quizzes, review tests, and examinations to give to our own faculty and any college or junior college teachers who are beginning to teach discrete mathematics. These are available upon request (for reproduction costs). Under the Sloan grant we have also prepared notes on "Network flows" and "Expert systems." We have taught "Network flows" for some time but that material was not in our original text [12]. Both sets of notes appear as chapters in the second edition of Mott-Kandel-Baker.

FEEDBACK.
We have discovered that the discrete mathematics students need rapid feedback. We try to accomplish that in several ways. Personally, I assign problems that have solutions in our text on one day, the next day I assign problems over the same material without answers, and then we discuss these in class. Occasionally, I may take up and grade one problem that was not discussed in class. Then the third day, I give a short ten minute quiz over that material. Throughout the term I give about eight weekly quizzes, two examinations, and a final. We encourage all of our teachers to give frequent quizzes. For major tests, we suggest that our teachers hand out a review test a week in advance. Personally, I discuss the problems from the review test that the students cannot work themselves. (Usually I ask for other students to give their solutions.) I post solutions to the review test. Then I give a test more or less following the same design. When I say the same design, I mean that, if on the review test, for example, problem 1 was an induction problem, problem 2 was a problem to fill in the blanks about properties of graphs, problem 3 had several parts to prove or disprove, problem 4 was a combinatorics problem, and problem 5 was a demonstration problem about isomophisms of graphs, then on the actual test I try to make problem 1 a different induction problem, problem 2 a fill in the blanks problem about properties of graphs, and so on. If the

combinatorics problem on the review test was, say, to count the number of 13-card bridge hands that has at least one card of each suit, then I generally would give a combinatorics problem on the actual test that also requires the principle of inclusion-exclusion. But, in my opinion, only a change in the number of cards is not a significant enough change in the problem. I might ask rather the following questions: How many 10-card hands have at least one honor card?

CHARACTERISTICS OF OUR PROGRAM.
The first characteristic of our curriculum is that our discrete sequence and our calculus sequence may be taken concurrently. They are not integrated into one sequence. In my opinion, this will be the predominant pattern suitable for the most people for the longest period of time and, morever, it requires the least major overhauls in our present curriculum. Besides that, trying to write materials for an integrated course to fit the needs of many different universities is a most uncertain task.

One characteristic of our course at FSU is that we put an emphasis on techniques of proof. On every major examination in Discrete I and Discrete II we encourage our teachers to ask for demonstrations that require induction, contrapositive, contradiction, negations of quantified sentences, and counterexamples.

When I say applications of induction, I don't mean just to verify some simple formula by induction. Instead, I attempt to ask for proofs by induction that require the ideas of the chapter we are discussing.

For instance, early in the term I may ask questions like:

(1) Prove $4^n > n^4$ for each integer $n \geq 5$.
(2) Prove that, for each positive integer n, a set with n elements has 2^n subsets.
(3) Given a list L of 2^n numbers arranged in increasing order, show that it is possible to determine whether a particular number k is in the list by comparing k to at most n+1 numbers in the list L.

But, then, in the chapter on combinatorics, I might ask:

(4) Prove that one half of the 6^n outcomes of tossing n distinguishable dice have an even sum.
(5) Prove that for all integers $n \geq 1$

$$\sum_{k=1}^{n} \binom{k+3}{4} = \binom{n+4}{5}$$

(6) Prove that there are m^n functions from a set A with n elements to a set B with m elements.
(7) Prove that for each integer $n \geq 1$ the Stirling numbers of the second kind satisfy the following:

a) $S(n,2) = 2^{n-1} - 1$
b) $S(n, n-1) = C(n, 2)$.

Then in the chapter on graphs:

(8) Prove that every path of length n contains a simple path. Conclude that a circuit always contains a cycle.
(9) Prove that, for each integer $n \geq 2$, if $d_1, d_2, \ldots, d_n$ are positive integers such that

$$\sum_{k=1}^{n} d_k = 2(n-1)$$

then there is a tree T_n with n vertices whose degrees are $d_1, d_2, \ldots, d_n$.
(10) Prove Euler's Formula for plane connected graphs by induction on
a) the number of cycles.
b) the number of edges.
(11) Prove, if R is a symmetric relation, then R^n is symmetric for each integer $n \geq 1$.
(12) Prove that the vertex chromatic number for the complete k-partite graph $k_{n_1, n_2, \ldots, n_k}$ is k.

Induction is particularly useful when considering strings over a finite alphabet. Here are two simple applications that can be assigned almost any time in the term.

(13) A *palindrome* can be defined as a string that reads the same forward and backward, or by the following definition.
a) The empty string is a palindrome.
b) If a is any symbol, then the string a is a palindrome.
c) If a is any symbol and x is a palindrome, then axa is a palindrome.
d) Nothing is a palindrome unless it follows from (a) through (c).

Prove by induction that the two definitions are equivalent.

(14) The strings of balanced parentheses over the alphabet $\{(,)\}$ can be defined in at least two ways.
a) A string w is balanced if and only if:

1. w has an equal number of ('s and)'s, and

2. any prefix of w has at least as many ('s as)'s.

b) 1. The empty string is balanced.

2. If w is a balanced string, then (w) is balanced.

3. If w and x are balanced strings, then so is wx.

4. Nothing else is a balanced string.

Prove by induction on the length of a string that definitions (a) and (b) define the same class of strings.

At Florida State all computer science majors are required to take Discrete I and II while mathematics majors must take only the first term. The students are not required to have had either of our first programming courses (Pascal I and II) nor are they required to have calculus. When we first started teaching discrete mathematics, most students took Pascal I concurrently with discrete. That practice seems to have changed, however, for a random sample of two sections of Discrete I and II in the fall of 1984 and in the spring of 1985 reveals the following data:

	Calculus	Pascal I	Pascal II
94 students in two sections of Discrete I			
# of students having passed previously	45	32	7
# of students taking Discrete I concurrently and passing	7	17	9
129 students in two sections of Discrete II			
# of students having passed previously	87	63	48
# of students taking Discrete II concurrently and passing	3	2	37

Figure 2.

Thus, while our students are not required to take other courses as prerequisites, a fair percentage of them are choosing, nevertheless, to do so.

THE DEBATE GOES ON: TO COUNT OR NOT TO COUNT. Needless to say, not every one will agree that all topics in discrete mathematics are equally important. Our own faculty is not in complete agreement as to how much combinatorics and recurrence relations should be covered and when. Additonally, it is clear from the texts that are coming out that there is a difference of opinion nationally. For instance, J. Gersting's book [4] has three pages covering only simple combinations and permutations with five examples. Gersting has nothing on recurrence relations. Ross and Wright [15] have 20 pages and 10 examples. They have additional material in their chapter on counting about the pigeonhole principle and countable sets. Again they have nothing on solutions of recurrence relations.

On the other hand, R. Johnsonbaugh [7] has 52 pages divided about equally on combinatorics and solutions of recurrence relations with 35 examples. Mott-Kandel-Baker [12] has 106 pages on combinatorics covering both simple permutations and combinations as well as allowing repetitions and also covering the principle of inclusion-exclusion. For simple combinations and permutations alone, Mott-Kandel-Baker has 33 pages and 30 examples. But then [12] also has a chapter of 88 pages on solutions of recurrence relations with 45 examples.

Of course, applied combinatorics texts like Buraldi [1], Liu [10], Tucker [18], and Roberts [14] generally contain even more on combinatorics and all except Brualdi contain chapters on Polya Enumeration.

AUDIENCES THAT NEED DISCRETE MATHEMATICS. There are good reasons for the different opinions discussed above, not the least of which is the audience for whom they are written. There are three main audiences that need discrete mathematics: (1) computer engineering majors, (2) computer science majors, and (3) mathematics majors. Each of these audiences have similar but separate needs. Some of the texts are written (irrespective of their titles) with the specific needs of individual audiences in mind and some are written for more than one audience.

I would like to put forth a reason for including combinatorics in a discrete mathematics course aside from any specific needs in later courses. Unquestionably, problem solving is a skill we would like our majors to have. I believe that this is a learned skill that grows gradually from successful experience. There are several subjects that are conducive to learning this skill and combinatorics is one of those subjects. This, to me, is reason enough to include combinatorics.

APPRECIATION: In closing, I would like to thank the Sloan Foundation for all they have done to encourage the debate over discrete mathematics in the curriculum.

REFERENCES:

1. Brualdi, R. A., "Introductory Combinatorics", North-Holland: New York, 1977.

2. Chartrand, G., "Graphs as Mathematical Mod-

els", Prindle, Weber, and Schmidt: Boston, 1977.

3. Fulkerson, D. R., "Studies in Graph Theory, I and II", Studies in Mathematics, MAA.

4. Gersting, J., "Mathematical Structures for Computer Science", W. H. Freeman: New York, 1982.

5. Herstein, I., and Kaplansky, I., "Matters Mathematical", Harper and Row: New York, 1974.

6. Honsberger, R., "Mathematical Gems, I and II", "Mathematical Morsels", "Mathematical Plums", The Dolciani Mathematical Expositions, MAA.

7. Johnsonbaugh, R., "Discrete Mathematics", Macmillan: New York, 1984.

8. Kolman, B., and Busby, R., "Discrete Mathematical Structures for Computer Science", Prentice-Hall: Englewood Cliffs, NJ, 1984.

9. Larson, L., "Problem-Solving through Problems", Springer-Verlag: New York, 1983.

10. Liu, C. L., "Introduction to Combinatorial Mathematics", McGraw-Hill: New York, 1968.

11. Niven, I., "Mathematics of Choice", New Mathematical Library 15, MAA.

12. Mott, J., Kandel, A., and Baker, T., "Discrete Mathematics for Computer Scientists", Reston: Reston, VA, 1983. Second edition, Prentice-Hall: Englewood Cliffs, NJ, 1986.

13. Polya, G., Tarjan, R., and Woods, D., "Notes on Introductory Combinatorics", Birkhauser: Boston, 1983.

14. Roberts, F., "Applied Combinatorics", Prentice-Hall: Englewood Cliffs, NJ, 1984.

15. Ross, K., and Wright, C., "Discrete Mathematics", Prentice-Hall: Englewood Cliffs, NJ, 1985.

16. Stein, S., "Mathematics, the Man-Made Universe", W. H. Freeman: San Francisco, 1969.

17. Trudeau, R., "Dots and Lines", The Kent State University Press, 1976.

18. Tucker, A., "Applied Combinatorics", John Wiley: New York, 1980.

"COUNTING" VS "PROOFS"

Steven F. Bellenot

Perhaps the hardest goal of discrete mathematics is that of fostering mathematical maturity. Indeed, this maturity itself is hard to define. Somewhere along the line a mathematician learns how to do proofs. Before that time he or she isn't a mathematician but afterwards is or can be. There seems to be a belief in the threshold theory. That is, suddenly one sees the light or one is always left in the darkness. On the other hand, it is known that calculus level or discrete math students are not ready for the "prove or die" test. Are there stepping stones which will at least lower this threshold?

We will consider two topics in discrete math and test their effectiveness in achieving this goal. This will be a historical perspective from the author's teaching experience with discrete math. The appendices contain much of the information that remains from the 4 discrete math courses the author has taught. The two topics considered are "counting" (i.e. combination, permutation, selections, etc.) and "proofs".

Whereas "counting" isn't the same thing as "proofs", it can certainly be quite hard and knowing if you got the correct answer requires maturity. Anyway, my first discrete math course (Appendix 1) was in Fall 1981. We had a new text (Tucker [2]) and roughly the first four weeks were on induction and counting. (Even more counting came later.) I had been warned how badly the students do in this class, but I was surprised at how badly they did on the first test. (Appendix 1B has this test and a breakdown of how many students scored how much on each problem.) A number of quiz problems were like the problems on the test but the test didn't separate this kind of problem from that kind of problem. For example Problem 6 allows repetition but many people gave the answer for no repetitions.

(Actually some students were clearly over their head in this class. One homework problem was about 3 people each ranking 10 other people and asked how likely it was that person A was in the top 3 of at least 2 of the 3 rankings. One student asked if order was important. He decided the answer was either $10 \cdot 9 \cdot 8$ (permutation) or 10 choose 3 (combination). Another asked which number 10, 3, or 2 didn't matter.)

Well, what had gone wrong? On the final, problems were grouped better. During the last class before the final, we all said "at least one means not none" three times. Then came the next surprise; what little they knew about counting was forgotten. (Counting problems from the Final are in Appendix 1C. Unfortunately no scores. Note Inclusion-Exclusion is included.) Before giving grades I looked carefully at these finals and noted that the students who got D's almost all decreased their scores on counting. (See Appendix 1A.) Indeed, if counting wasn't a topic in this class, they all might have gotten C's. On the other hand there was a B student whose algebra was so bad I wouldn't have given him/her a C in Calculus.

From this first class it was clear that "counting" wasn't easy to teach. In some senses it requires more maturity to know why an answer is wrong than to get the right answer. For example, counting 5 card hands with at least one spade, one often gets the expression on the left

$$\binom{13}{1}\binom{51}{4} \qquad \binom{52}{5} - \binom{39}{5}$$

as an answer. It is much easier to explain why the expression on the right is correct than that the other is wrong. Furthermore, since there is often more than one way to get the correct answer, students are less likely to check if their answer is correct.

Eventually I did have better success in teaching counting (Appendix 4) but that is jumping the gun a bit. Also I currently believe that counting doesn't belong in discrete. But at the end of this first course I was ready to teach it again with grand plans on how to get counting across so that the students could understand it.

We now march into Spring 1982 and it is the second semester of discrete. This is the first time a second semester has been added. The topics are graph theory and proofs. for some reason (maybe because the class was small) I have a breakdown of their scores on each of the 3 hour tests. Fortunately there was a good deal of spare time in this class. I made each of the students go up to the blackboard and present a proof. I learned many of their simple errors could be removed with an understanding of Venn diagrams (hence Question 2 on Test 3 (Appendix 2)). Secondly, I learned the value of fully stating algorithms. DFS and BFS yield too many trees if you don't look at the vertices in some order.

This class had a rough time with proofs. Inductive proofs seemed to give them the most trouble. My experience with other students passing Discrete 1 seems to indicate that most could do simple induction proofs. But we notice that examples and counterexamples are doable (i.e. the relation in Problem 4, test 2).

Fall of 1982 and another Discrete 2 (finally discrete was named discrete). This class was very good and many could handle the simple proofs given to them. Perhaps ironically the weakness of this class was lack of algebra. Recurrence relations were now in Discrete 2 and scores on that section were lower than the others. The first week of class was spent on Venn diagrams and valid and invalid conclusions.. Counterexamples are doable (i.e. the relation in Problem 4, test 2).

It was a year later before I taught another discrete course. I was teaching computer science courses, but in Fall 1983 I was "Duke of Discrete". I planned the

course outline for the new text (Mott, Kandel, Baker) and carefully omitted any "counting" from Discrete 1. However, I was overruled and "counting" is currently in Discrete 1. These issues will be discussed later.

Spring 1984 and back to Discrete 1. Again we spent an early week on Venn diagrams and valid and invalid reasoning. The students scored 94% on this question on the Final (Problem 5 (see Appendix 4)). We see questions like 8 asking for examples of graphs rather than the definitions asked for in Appendix 2 (with 76% vs 100%, 80%, 77%, 71%, 66%). Note the counterexample problems (17 and 18) and how well it agrees with the final averages at both ends. (Counterexample problems are easy to grade. It is either right or wrong and usually the students missing the problem are the weaker students.) The relation problem seems to be much harder than the one in Appendix 2, Test 2. Note also that the proofs go down as one would expect. The low showing on the induction proof (Problem 20) is again due to lack of algebra.

Well, what about counting? The A, B, C and D's scored almost exactly their average on these problems whereas the F's are fully 17% below their average on these problems. Why is this so different from the first discrete class? Note that, with the exception of the inclusion-exclusion questions, the two finals have nearly identical counting problems. Two things were done differently. There were harder problems done in class but about half as many. Two, counting was done at the end so they wouldn't have time to forget it. Note the improvements in the third test to final in Appendix 4A vs the first test to final in Appendix 1A. Notice also how counting problems were harder than proofs on the final.

Conclusions.

Proofs can be taught in a Discrete 1 course if you spend time on the stepping stones of Venn diagrams, examples and counterexamples. Counting should only be taught near the end of the class if at all.

Actually counting doesn't belong in discrete math at all. Except for Discrete 2 the counting needs of the other courses requiring Discrete 1 are simple. Digital Networks simply notes that there are 2 to the power 2^n switching functions of n variables. Programming 2 needs to know there are $n!$ permutations of $1 \ldots n$.

The goal of mathematical maturity is best met with stepping stones to proofs. Adding the confusion of counting to calculus-level students is perhaps asking too much. In any case counting knowledge is soon lost due to lack of use.

As a final note, here are the reasons "counting" was put back into Discrete 1: (1) "counting" is a traditional part of what mathematicians think of as discrete math. (2) Only the first semester of discrete is required of math majors (pure not applied). (3) This Sloan grant is for developing discrete math for math majors as well as for computer scientists. (4) Chapter 2 of the current text, written "in house", is on counting. (5) Students and many professors hate to skip around in a text.

Perhaps the most damning of reasons not to teach counting is the mathematicians' eagerness to teach it. This eagerness isn't just because it is good mathematics (which it is) but because the professor will finally get a chance to learn it. Something the mathematician has not found time to do otherwise.

Someone should write a discrete text with no chapter on counting.

TEXTS

1. Mott, J.A., Kandel and T. Baker, Discrete Mathematics for Computer Scientists, Second edition, Prentice-Hall, Englewood Cliffs, NJ, 1986.

2. Tucker, A., Applied Combinatorics, Wiley, New York, 1980.

Appendix 1–Discrete Math 1–Fall 1981

A. Grades

Grade	A	B	C	D	F	Drops
Number of Students	0	4	1	6	17	8

Only 8 of the F's took the final.

Scores: first test, final (percentage)

B's: (83, 81) (84, 70) (63, 80) (76, 84)

C: (56, 63)

D's: (78, 70) (75, 60) (69, 72) (63, 58) (53, 44) (50, 50)

F's: (68, 48) (74, 66) (65, 49) (52, 55) (54, 46) (43, 32) (57, 48)

The big loss seemed to be in counting. In general, they seemed worst at counting on the final than at the time of the first test.

B. First test—all problems but last worth 5 pts. Score $(x, y, z, a, b, d) = x$ got 5 pts, y got 4 pts., etc.

1. How many arrangements are there of the word arrangements? Score (20,0,1,0,0,5); percentage: 79%

2. President Reagan wants to give you nine jelly beans of the same color. (It was to be ten, but budget cutbacks, you know.) On his desk he has a large paper bag full of jelly beans in 13 different colors. What is the *fewest* number of jelly beans old Ron needs to take out of the bag always to have enough of one color to give to you? Score (15,0,0,0,0,11); percentage: 58%

3. A and B are sets. A has 6909 elements, B has 1107 elements and $A \cap B$ has 225 elements. How many elements does $A \cup B$ have? Score (23,0,0,0,0,3); percentage: 88%

4. How many license plates are there starting with either '#' or '*' or a blank followed by 2 letters, then 3 digits and ending with either a letter or digit or '$' or '?' or blank? (There was a picture with this problem pointing to the positions.) Score (11,0,0,8,2,5); percentage: 56%

5. What is the probability that a five card poker hand has the jack of diamonds, the seven and three of spades, the ace of clubs and the six of hearts? Score (13,0,0,4,7,2); percentage: 62%

6. How many ways are there to select 23 ice cream cones from 31 flavors? Score (8,0,0,3,0,15); percentage: 35%

7. A group of 10 men and 3 women want to form a committee with at least two people which has twice as many men as women. How many ways can they do this? Score (20,0,0,2,0,4); percentage: 80%

8,9,10. How many 4 card hands (from a deck of 52) have:

8. Exactly one card of each suit? Score (10,0,0,0,4,12); percentage: 42%

9. Exactly one pair? Score (9,0,1,6,0,10); percentage: 46%

10. At least one pair? Score (8,0,0,1,4,13); percentage: 35%

11. How many ways are there to put 17 identical pennies into 6 different parking meters so that exactly two of the parking meters get none? Score (7,0,3,3,2,11); percentage: 40%

12. How many ways are there to roll 101 *distinct* dice so that *exactly* 37 "three's" are rolled? Score (11,0,0,2,3,10); percentage: 48%

13. How many non-negative integer solutions are there to $x_1 + x_2 + x_3 \leq 10$? Score (8,0,0,2,2,14); percentage: 35%

14,15 How many arrangements of "finitemath" are there with

14. No consecutive vowels? Score (1,0,0,2,5,18); percentage: 11%

15. With the 'F' and 'M' next to each other? Score (4,0,0,7,1,14); percentage: 27%

16. There are 3 different roads from A to B and 4 different roads from B to C. There is also one road from A to C which by-passes B. How many

ways can you go from A to C and back to A so that you do not repeat any road (or portion thereof) on the way back? Score (5,0,2,1,0,18); percentage: 25%

17. (Worth 20 pts.) Prove by induction:

$$\sum_{i=1}^{n}(2i-1) = n^2$$

Score (11,5,2,0,0,2,1,0,0,0,0,0,4,0,0,1,0,0,0,0); percentage: 82%

C. Final questions on counting. Unfortunately no scores. (20 points 1-4, 10 points for 5 and 6)

1. How many license plates are there with:
   A. Three letters followed by three digits?
   B. With either two, three or five letters?
   C. With 8 digits, exactly 5 of them zero's?
   D. With 6 letters, exactly 2 of them vowels?

2. How many arrangements of "MerryChristmas" are there?
   A. Altogether?
   B. With the 'Y' trapped between the 'M's' (no letter between)?
   C. With the 'Y' next to the 'C'?
   D. With no consecutive vowels?

3. How many ways are there to pick 20 electronic games from one of Santa's 50 kinds with:
   A. No repetition?
   B. Unlimited repetition?
   C, D. (Worth double) Use Inclusion-Exclusion to find the number of ways you can pick up to five games of the same kind.

4. What is the probability that a 5-card hand (from a deck of 52) has:
   A. All face cards ($A, K, Q, J, 10$, are face cards)?
   B. At least one red card (hearts and diamonds are red)?
   C. Exactly one pair (no three or four of a kind)?
   D. At least one pair (or more)?

5. How many ways are there to roll 100 distinct dice with at least one of each kind of face showing?

6. There are 4 roads from success to scandal and 10 roads from scandal to ruin. Also there are 5 roads from success to ruin which avoid scandal.
   A. (2 points) How many roads to ruin (routes from success) are there?
   B, C, D. (8 points) How many ways are there to go from success to ruin and back that doesn't repeat any portion of the route there on the way back?

Appendix 2–Discrete Math 2–Spring 1982

A. Grades

Grade	A	B	C	D	F
Number of Students	2	1	5	2	4

Two of the Fs took final, one stopped at the third test, one quit after the first test.

Selected test questions each worth 10 pts. Score $(x, \ldots z)$ again x got full credit, z got zero points.

Test 1
1. Define:
   A. degree of a vertex–Score (14,0,0,0,0,0); %100
   B. coloring of a graph–Score (8,2,0,4,0,0); %80

Test 2
1. Define:
   A. a tree   B. a backedge
   Score (1,2,2,3,3,1,1,0,0,0,0); %71

4. Let R be the relation on the real numbers where xRy if and only if $x - y$ is a nonnegative integer.
   A. Is R reflexive?
   B. Symmetric?
   C. Transitive?
   D. Anti-symmetric?
   Score (8,0,0,3,0,5,0–); %85

Test 3
1. A. In a network what is $k(P, \overline{P})$? (Explain P and $\overline{P}$ as well.)
   Score (2,3,6,1,1,0); %66
   B. Explain the difference between a tree and a rooted tree.
   Score (4,3,6,0,0,0); %77
2. A. Draw a Venn diagram for each of the three statements
   (1) Each A is B
   (2) All C is B
   (3) Each A is C
   Score (4,4,4,0,1,0); %75
   B. Is it logically valid?
   Score (11,0,0,1,1,0); %89

Appendix 3–Discrete Math 2–Fall 1982

A. Grades

Grade	A	B	C	D	F	Drops
Number of Students	5	4	2	2	5	3

No failing student took the 3rd test or the final

Appendix 4–Discrete Math I–Spring 1984

A. Grades

Grade	A	B	C	D	F	Drops
Number of Students	1	3	4	1	16	11

Only 6 of the F's took the final.
- A: (86,93) gain 7%
- B: (76,84) (84,85) (84,88) avg. gain 4%
- C: (70,75) (80,81) (67,82) (63,69) avg. gain 7%
- D: (55,72) gain 17%
- F: (52,54) (52,54) (46,40) (55,48) (73,39) (62,61) (43, 50) avg. loss 6% (but only 1% if the one out of line is omitted)

B. Final questions, score, percent (note graphs are omitted)

1. For graph 1 find a spanning tree by
   A. DFS
   B. BFS
   C. List the vertices of graph 2 in postorder.
   Score (7,3,2,2,0,0,1,1,0,0,0); 84%

2. A. Draw all (undirected) trees with 5 edges. No two trees in your list can be isomorphic.
   B. Negate "there is a y so that, for each $x, y > 0$ and $y \neq x^2$" so that your answer does not contain the word "not".
   Score (2,3,0,5,1,1,3,1,0,0,0); 68%

3. A. Change $(a + b)(c + d/e) + f$ to prefix
   B. Build the BST for 40,60,20,45,30,25,90,100,70, 50,75
   Score (7,1,1,3,2,2,0); 81%

4. A. Find GCD (1462,123) by the Euclidean Algorithm (show work!)
   B. In Z_{13} find i with $0 \leq i \leq 13$ so that $[i] = [2]/[19]$.
   Score (6,1,0,2,2,2,0,0,1,0,2); 67%

5. Draw a Venn diagram for A and B
   A. All x is y
   B. No z is x
   C. For each of the statements below label the statement a valid or invalid consequence of statements A and B above. (1) no y is z; (2) no x is z; (3) some y is z.
   Score (11,4,0,0,0,0,1,0,0,0,0); 94%

6. In the direct tree T every parent has exactly 4 children. T has 100 leaves.
   A. How many nodes does T have?
   B. How high must T be?
   C. How high could T be?
   Score (6,0,3,2,1,0,2,0,1,0,1); 71%

7. Either produce an isomorphism or show why none exists.

A. Between graph 3 and graph 4.
B. Between graph 4 and graph 5.
C. Between graph 3 and graph 5.
Score (10,0,0,1,1,0,1,2,0,1,0); 78%

8. Give examples.
   A. Two non-isomorphic loop-free graphs with degree sequence (1,1,2,2,3,3)
   B. A digraph of a non-reflexive relation which isn't irreflexive either.
   C. A maximal disconnected subgraph of graph 6 with 5 edges.
   Score (6,0,0,7,0,0,3,0,0,0,0); 76%

9. Define R on $\{1, 2, 3, \ldots\}$ by xRy iff $x(y + 1)$. For each property below either say no and give a counterexample or say yes.
   A. Reflexive?
   B. Irreflexive?
   C. Transitive?
   D. Anti-symmetric?
   Score (0,0,0,5,0,2,1,3,1,4,0); 40%

10. For the Hasse diagram of the partial order R given by graph 7
    A. list all maximal elements
    B. list all minimal elements
    C. list all elements "less than or equal to d"
    D. Find $\inf(b, g)$
    E. Find $\sup(j, h)$
    Score (3,3,4,4,0,0,1,1,0,0,0); 78%

11. How many 5 card hands have
    A. Exactly 3 spades and exactly one club?
    Score (8,1,2,4,0,1); 73%
    B. At least one spade?
    Score (6,0,1,2,1,6); 48%

12. What is the probability of an arrangement of $ABBCDEEFG$
    A. Having the C next to the D?
    Score (10,9,2,3,1,0); 79%
    B. Having no consecutive vowels?
    Score (2,2,2,6,3,1,); 49%

13,14. How many ways can you put 100 balls into 239 distinct boxes?

13. A. If the balls are distinct and repetition is allowed?
    Score (8,1,0,3,3,1); 66%
    B. If the balls are identical and no repetition is allowed?
    Score (11,0,0,0,2,3); 71%

14. A. If the balls are identical and repetition is allowed?
    Score (8,2,1,2,2,1); 71%

B. If the balls are identical and no box can have more than 75 balls?
Score (0,2,0,3,0,11); 18%

15. How many 4-member committees can be formed from a group of 5 French, 6 Cubans and 4 English

  A. With exactly one French person?
  Score (10,0,0,3,1,2); 71%

  B. With at least as many French as Cubans and at least as many Cubans as English?
  Score (8,2,1,0,1,4); 65%

16. Prove:

  A minimal connected subgraph containing the nodes x, y and z of the connected graph G is a tree.
  Score (5,0,3,0,2,4,2,0); 71%

17, 18. Give counterexamples

  A. Every connected graph with a cut-edge has a cut- node.

  B. A tree with 5 or more edges has either at least 4 vertices of degree 1 or all vertices of degree $\leq$ 2.

  C. An equivalence relation on a set with at least 2 elements is never a partial order.

  D. If $f(n)$ and $g(n)$ are positive functions for all $n \geq 1$ and $0(f) = 0(n)$ and $0(g) = 0(n^2)$, then $f(n) \leq g(n)$ for all $n > 1$

  E. In any binary tree with 3 or more nodes, pre-order is always different from inorder. Score (3 got 20, 1 got 17, 7 got 13, 4 got 9, 1 got 1); 65%

19. Prove. If G is a connected loop-free graph which is not a tree, then G has at least two spanning trees. Score (0,0,3,4,1,4,0,0,1,0,3); 50%

20. Given $a_0 = 0$ and $a_n = 2a_{n-1} + 2^{n+2}$ for $n \geq 1$. Prove by induction $a_n = 4n2^n$ for $n \leq 0$. Score (6,2,0,1,1,1,0,4,0,1,0); 65%

C. Final grade vs problem scores

	Final Grade (No.)		
	A&B(4)	C&D(6)	F(6)
Prob. No.			
11A	80%	83%	50%
11B	95%	50%	20%
12A	100%	93%	57%
12B	80%	50%	20%
counting 13A	65%	93%	33%
13B	100%	100%	23%
14A	95%	80%	47%
14B	75%	7%	0%
15A	100%	90%	43%
15B	95%	80%	27%
counting average	88%	73%	32%
examples/ 8	93%	75%	65%
counterexamples 9	50%	35%	38%
17 & 18	88%	63%	52%
example average	79%	60%	52%
16	95%	73%	53%
proofs 19	78%	53%	32%
20	98%	80%	37%
proofs average	88%	69%	41%
example/proofs average	83%	64%	47%
final average	88%	74%	49%

DISCRETE MATHEMATICS I
AT
TALLAHASSEE COMMUNITY COLLEGE
Joseph F. Hoffmann and Bettye Anne Case

Due to the state-wide articulation agreements between public two-year colleges and universities, Florida State University cannot require a course as an entrance requirement at the junior level until it is offered by a substantial number of the two-year colleges. A successful model of a Discrete I course within the state at the two-year level was needed to demonstrate the feasibility of the inclusion of discrete mathematics as an offering in the same way the calculus is included in the curriculum. The funding from this grant allowed Hoffman release from another class in Spring 1984 to teach the course at Tallahassee Community College (T.C.C.). T.C.C. is a small college; for example, Calculus I has only one or two sections per semester. It is thus intended to include Discrete I each spring semester only. The course was included as a regular offering in 1985 and 1986 and we describe below Hoffman's reporting on the 1984 and 1985 experience. Due to the state's common numbering system and the close articulation relationship already existing between T.C.C. and F.S.U., matters of topic inclusion and credit acceptability were smoothed. The experience is now being studied by St. Petersburg, Chipola and other two-year colleges in the state and it is hoped they will soon include the course.

As it worked out, Hoffman actually first taught Discrete I at F.S.U. during Fall 1983 as an adjunct instructor. During Spring 1984 he taught the project-funded T.C.C. section and a section at F.S.U. Comparisons between his methods and sections are therefore appropriate. The course and topics were essentially the same, but Hoffmann notes these similarities and differences:

1. At T.C.C. the spring semester of 1984 was one week longer than at F.S.U.; consequently, with my T.C.C. class I had three additional 50-minute meetings, a fact that allowed me to move at a slightly slower pace and go over a few more exercises from the textbook in class.

2. The somewhat smaller size of my T.C.C. class (17 vs. 31 students) gave my students there a small advantage in the personal attention I gave them and in the slightly easier access to asking questions they had.

3. At T.C.C. I gave four 55-minute tests, whereas at F.S.U. the students took three 80-minute tests. Hence, my T.C.C. students were tested over smaller blocks of material at a time.

4. At both schools, there was a two-hour comprehensive final examination; these final exams were virtually identical.

Statistical comparisons of Hoffmann's Spring 1984 and 1985 classes at T.C.C. and the Spring 1984 class at F.S.U. in tabular form follow.

Discrete Mathematics I

	F.S.U. Spring 1984	T.C.C. Spring 1984	T.C.C. Spring 1984
Initial enrollment	31	17	11
Student background			
College Algebra level only	10	10	7
Calculus I or more	21	7	4
Intended majors			
Computer Science or Mathematics	27	13	8
Electrical Engineering	1	1	1
Mathematics Education	1	1	1
Undecided	2	2	1
Final grades (parentheses indicate number who had completed Calculus I)			
A	3(3)	3(1)	1(1)
B	5(4)	1(1)	1(1)
C	9(8)	9(5)	2(1)
D	7(3)	1(0)	3(1)
F	7(3)	0	2(0)
W	0	3(0)*	1(0)*

*At T.C.C., students were allowed to withdraw from a course up through the last day of classes in the semester. (This policy is no longer in effect.)

A STUDY OF PERFORMANCE ON A FINAL EXAMINATION IN DISCRETE MATHEMATICS II

Ralph D. McWilliams

The following is an informal analysis of performance on the final examination by the forty students in a class in MAD 3105, Discrete Mathematics II, in the Spring Semester 1984 at Florida State University. In Part I, the 13 problems are listed, along with a breakdown as to the types of errors students made on the individual problems. In Part II, overall achievement on the final examination and in the course is compared with success in previous mathematics courses taken. Each part has a summary stating some general observations.

Analysis of Performance on Each Problem of Final Examination

Problem 1.

> In how many ways can three cards be selected from a deck of 52, without replacement, if the first is a heart, the second is a seven, and the third is not a three? Show analysis of the problem, and give the answer in terms of sums and products.

Average credit: 3.625 points out of possible 6.

Of the class of 40,

13 did the problem correctly;

5 analyzed it correctly, but made a minor error along the way;

9 had a good start but then made a logical error;

13 had a completely wrong analysis of the problem.

Specifically,

8 had confusion with the fundamental principles of addition and multiplication (i.e., with "and" and "or");

5 had difficulty with the concept of combinations;

5 made elementary errors in counting.

Problem 2.

> A multiple-choice test has 20 questions and 5 choices for each answer. In how many ways can the questions be answered, so that each question is answered and at least 17 of the answers are correct?

Average credit: 3.1 points out of possible 5.

Of the class of 40,

    6 did the problem correctly;

    9 did it correctly except for a minor error;

    6 greatly oversimplified the problem;

    9 overlooked overlapping classes of outcomes (i.e., overlooked the fact that certain classes of outcomes were not mutually exclusive);

    10 had analyses of the problem that were essentially totally wrong.

Specifically,

    6 confused the significance of expressions such as 2^4 and 4^2 in counting outcomes;

    3 had trouble with "and" and "or";

    2 found exactly the number of outcomes not of the required type.

Problem 3.

In how many ways can 50 identical balls be placed in 6 distinct boxes if exactly 2 boxes are empty?

Average credit: 3.64 points out of possible 6.

Of the class of 40,

    6 were correct;

    19 after selecting the 2 empty boxes, failed to insure rest were empty;

    1 failed to use the formula $C(n+r-1, n)$ correctly;

    10 made two or more errors of the above two types;

    4 had analyses that were essentially entirely wrong.

Problem 4.

In how many ways can 22 distinguishable books be given to 5 different students, so that 2 students get 6 books each, 2 students get 3 books each, and one student gets 4 books? (We are not concerned with the order in which a given student receives his/her books, but are concerned with which students get 6 books, etc., and with which particular books each student gets.)

Average credit: 3.86 points out of possible 6.

Of the class of 40,

    6 did the problem correctly;

    1 did it correctly except for a minor algebraic error;

    21 erred in determining how many books each student gets;

    4 erred in determining which books each student gets;

    8 had major errors in logical analysis

Problem 5.

By use of the binomial theorem, complete and prove:

$$\binom{18}{0} + 2\binom{18}{1} + 2^2\binom{18}{2} +$$

$$\cdots + 2^k\binom{18}{k} + \cdots + 2^{18}\binom{18}{18} = \underline{\quad}$$

Average credit: 3.23. points out of 4.

Of the class of 40,

    24 did the problem correctly;

    12 tried to use the binomial theorem, but used it in a way that didn't help;

    4 had no idea what to do.

Problem 6.

We have 3 girl scout troops of 16 girls each. No girl belongs to more than one troop. In how many ways can we choose a committee of 8 girls having at least one girl from each troop? Use Inclusion-Exclusion (no credit otherwise).

Average credit: 3.53 points out of 6.

Of the class of 40,

    11 did the problem correctly;

    1 did it correctly except for a trivial careless error;

    8 used Inclusion-Exclusion correctly but made a moderately serious error;

    14 tried to use Inclusion-Exclusion, but made a serious error;

    6 got essentially nowhere with the problem.

Problem 7.

The alphabet has 26 letters, of which 5 are vowels and 21 are consonants. Find a recurrence relation for a_n, the number of n-letter "words" not having 2 consecutive vowels. (Repetition is allowed.)

Average credit: 4.28 points out of 6.

Of the class of 40,

    16 did the problem correctly;

    4 obtained the recurrence relation except for one error with a coefficient;

    10 got it correct except for two or more errors with coefficients;

    10 had a fatal error in analyzing the problem.

Problem 8.

Find the coefficient of x^{50} in $(x^3 + x^4 + \cdots + x^8)^{10}$.

Average credit: 4.33 points out of possible 8.

Of the class of 40,

    5 did the problem correctly;

    2 did it correctly except for minor algebraic errors;

    4 made a good small first step but nothing more (factored out x^{30});

    4 made the above small first step, then oversimplied the finite series to an infinite series, and completed the job correctly for the oversimplied problem;

    18 got correctly to the point of expanding in series, but then could not complete the job;

7 got essentially nowhere.

Problem 9.

Find a generating function for a_n, the number of ways of selecting n coins from an unlimited number of pennies, an unlimited number of dimes, and 100 half-dollars, given that we must select an odd number of pennies and at least 10 half-dollars.

Average credit: 2.90 out of possible 4.

Of the class of 40,

20 did the problem correctly;

4 had essentially 2 of the 3 factors of the generating function;

10 had essentially 1 of the 3 factors of the generating function;

6 had essentially an entirely faulty analysis.

Problem 10.

Find a generating function $A(x)$ for the recurrence relation $a_n + 5a_{n-1} - 36a_{n-2} = 0$ for $n \geq 2$, given that $a_0 = 0$ and $a_1 = 2$. (Express $A(x)$ as a quotient of polynomials.)

Average credit: 4.95 points out of 6.

Of the class of 40,

15 did the problem correctly;

11 had only minor technical errors;

6 made good progress, but had about 1 serious error;

5 made good progress, but had about 2 serious errors;

3 found the series approximately correctly, but did little more.

Problem 11.

Solve the recurrence relation $a_n + 5a_{n-1} - 36a_{n-2} = 0$, for $n \geq 2$, given that $a_0 = 0$ and $a_1 = 2$. Use the characteristic polynomial.

Average credit: 4.65 points out of 6.

Of the class of 40,

24 did the problem correctly;

1 did it correctly except for a minor error.

4 erred in getting the roots of the characteristic polynomial,

5 got the correct roots but erred in getting the numerical coefficients;

6 got the roots but did essentially nothing more.

Problem 12.

Given the recurrence relation $a_n + 4a_{n-1} - 9a_{n-2} - 36a_{n-3} = f(n)$, and noting that $t^3 + 4t^2 - 9t - 36 = (t+4)(t+3)(t-3)$, give the form of a particular solution of the inhomogeneous equation, if: (no partial credit on the parts of #12)

(a) $f(n) = 5 \cdot 2^n$

(b) $f(n) = 5n^3 2^n$

(c) $f(n) = n^4$

(d) $f(n) = 1$

(e) $f(n) = 4^n + (-4)^n$

(f) $f(n) = (n^2 + 1) \cdot 3^n$

Average credit: 3.45 points out of possible 6.

For #12 we give a different type of analysis:

Part	(a)	(b)	(c)	(d)	(e)	(f)
Number of students getting part correct	30	21	21	32	16	19

Problem 13.

For the network flow shown, use the algorithm studied in class to find a maximal flow from a to z. No credit unless you use the algorithm. Draw a new diagram for each run through the algorithm. That is, do not simply cross out and rewrite numbers as you go from one run to the next. After each run through the algorithm, list the flow chain along which you are increasing the flow and give the new net flow from a to z after the run. Also, explain why the procedure finally ends.

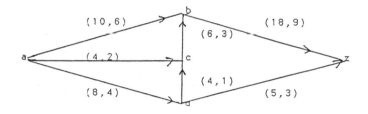

Average credit: 5.13 out of possible 6.

Of the class of 40,

28 did the problem correctly;

5 had a reasonably good understanding of the algorithm, but made errors in carrying it out,

6 had some understanding of the algorithm, but made serious errors, in some cases not realizing that in-flow and out-flow at a vertex must be equal;

1 got essentially nowhere.

Summary. The easier problems on the test were Problem 5 (an easy application of the binomial theorem), Problem 9 (finding a generating function for a selection problem), Problem 10 (finding a generating function for a recurrence relation), and Problem 13 (finding a maximal flow).

One of the most interesting situations is that concerning Problem 1, which requires very little formal knowledge of new concepts, but rather is largely a matter of "common sense." Trying to overcome difficulties of this kind is simultaneously a challenge and a frustration for the instructor.

The difficulty with Problem 8 is probably largely a matter of weakness in algebra. Also, additional prior contact with infinite series would probably make the students more comfortable with problems of this kind, but this hardly justifies increasing the prerequisite.

All the problems other than those mentioned in the first paragraph of the summary are deserving of increased concentration in classroom discussion.

Performance as Related to Background

To be eligible to take Discrete Mathematics II the student must have earned a grade of C or better in College Algebra and Trigonometry, MAC 1132 (4 semester hours), and a grade of C or better in Discrete Mathematics I, MAD 3104 (3 semester hours). Many students in Discrete Mathematics II have, of course, had various other courses, such as Calculus and Linear Al-

gebra. The system of prerequisites is strictly enforced.

The Department of Mathematics and the Department of Computer Science both have the policy that a student accumulating more than five grades of D or F in mathematics and/or computer science courses may not continue as a major. However, some students presently in the program have been in the university long enough so that they are not bound by that policy. Thus, some students taking Discrete Mathematics II may have considerably more than five grades of D or F in mathematics or computer science.

For purposes of this study the 40 students in the class have been placed in four categories, according to their background, and their performance on the final examination and in the course has been tabulated by category:

Category of Background	Students in Category	Grades on Final Examination	Grades in Course
Strong (mostly A and B grades in previous courses)	10	4A / 3B / 3C	3A / 4B / 3C
Fairly good (mostly C grades, with at most one D grade; perhaps some A and B grades)	16	2B / 11C / 2D / 1F	3B / 13C
Rather weak (few A or B grades, several C grades, at least 2D or F grades)	10	1B / 3C / 3D / 3F	7C / D / 1F
Very poor (numerous D or F grades)	4	3D / 1F	1C / 3D

Summary. It appears that the prerequisites for MAD 3105, Discrete Mathematics II, are appropriate as preparation for that course. Further, there is a good correlation between the student's background (i.e., grades in collge mathematics courses taken previously) and achievement in MAD 3105. This is especially true of the comparison of background with grades on the final examination. It may be noted that course grades are somewhat higher than final examination grades; this is probably due to some of the tests prior to the final examination's having been easier than they should have been. Grades on the final examination, however, seem quite consistent with the analysis of the students' backgrounds.

Broadly speaking, a student who has taken each of his or her other mathematics courses just once, with a grade of A, B, or C, is quite likely to receive an A, B, or C in MAD 3105. On the other hand, a student who has repeated one or more other courses is likely to do the same with MAD 3105.

Students repeating MAD 3105 received 3C's, 3D's, and 2F's on the final examination, and 5C's and 3D's in the course. This again indicates that a student who must repeat a course runs a substantial risk of again failing to achieve a satisfactory grade.

No trends were discerned that specifically related difficulties in MAD 3105 with grades in previous courses. Throughout the final examination, students with strong backgrounds tended to do well, while students with weak backgrounds tended to do poorly.

BACKGROUND AND PERFORMANCE OF 1700 STUDENTS IN DISCRETE MATHEMATICS
Academic Year Semesters:
Fall 1982–Spring 1985
Bettye Anne Case

By analysis of grades earned by students, we hope to find indicators which will lead to improvements in student achievement. No assumption is made that the student population is homogeneous either by section or by semester. There is wide variation in the grades awarded in the classes of various instructors in particular semesters, and between classes taught in different semesters by the same instructor. (Within one semester, the range in percent of D's and F's by section varies from 20 percent to 50 percent.)

All academic-year sections for six semesters were considered. The percentage of "satisfactory" (A, B, C) grades and of "unsatisfactory" (D, F) grades was computed for all those completing Discrete I and II. (Florida State has a very strict drop policy and so withdrawals may be neglected in analysis.) Several changes have taken place over this time. There was a text change after the first two of the six semesters, and small modifications each semester to test material, syllabus and course content. Mid-way through this time mathematics majors other than certain applied majors were required to take Discrete I. At the beginning of academic 1984-85, both computer science and mathematics became limited access degree programs. Perhaps that accounts for the slightly better grades than in previous terms for 1984-85 and it at least partially accounts for a drop in the number of students that year.

Instructors other than regular department faculty have been assigned to teach some Discrete I sections since the beginning of the Sloan project. (Teaching assistants writing doctoral dissertations, adjunct instructors with the doctorate in mathematics and full-time instructors holding a master's degree have taught Discrete I.) The differences in grades between these sections and sections taught by regular faculty are not greater than among faculty sections; there is a tendency for the non-faculty sections to have D and F percentages at the extremes of the distribution, although this seems to be

true only the first time the course is taught by such an instructor. It appears that upon second assignment to teach discrete mathematics, regular mathematics faculty and other instructors have similar grade distributions.

We summarize for Discrete I:

	Number of Students	Percent ABC	Percent DF
Fall 1982	190	61	39
Spring 1983	200	62	38
Fall 1983	216	60	40
Spring 1984	171	56	44
Fall 1984	151	62	38
Spring 1985	104	71	29
All Terms	1033	61	39

For Discrete II:

	Number of Students	Percent ABC	Percent DF
Fall 1982	88	59	41
Spring 1983	128	80	20
Fall 1983	97	77	23
Spring 1984	142	71	29
Fall 1984	93	63	37
Spring 1985	73	79	21
All Terms	621	72	28

The consistently high rate of failure in Discrete I and anecdotal difficulties of some transfer students prompted a closer analysis of the background vs. performance of all Fall 1984 Discrete I students. Six students withdrew prior to the end of the semester; of those, three had completed a calculus course (calculus grades: A,B,B) and three only a college-algebra level course (algebra grades: B,C,C). All but three students (who each received F's) appear to have at least the stated course prerequisite. Of the students for whom we have information, 88 took Discrete I after the F.S.U. College Algebra and Trigonometry course or a college algebra judged at a "high" level from another school. Thirty-four students had completed Calculus I or a higher course prior to Discrete; four were admitted as bachelor's degree holders (grades: A,A,A,B) and four were placed into the course by a test based on the MAA Calculus Readiness and Advanced Algebra tests (grades: A,A,B,C).

The overall grade distribution for Discrete I with breakdown for the "college algebra only" and "calculus I or equivalent" background groups follows:

Percents of Letter Grades Earned
Related to Student Background

	All Students	College Algebra Only	Calculus I or Equiv.
A	10	3	21
B	19	6	29
C	33	35	29
D	13	16	12
F	25	30	9

The drastic difference between the distributions grouped by background was not expected, but the percentage of D and F for the college-algebra-background students, 46 percent, is very close to the F.S.U. 45 percent D and F rate for Calculus I. Looking separately at students having their last previous course at Florida State and those transferring all previous courses, there is not a lot of difference in the success rates of the continuing and transfer students. Mean letter grade decrease from the last preceding course for continuing students having only college algebra was -1.4 and for transfers -1.6. Students with a calculus course or higher, continuing and transfers, also averaged a drop in grade for discrete: the continuing students ($-.45$ letter grade) may be significantly stronger than transfers ($-.71$ letter grade).

The difficulty of students taking Discrete I with the minimal advised prerequisite is being addressed, although some differences due to the "maturity factor" are probably inevitable. Since calculus content is not used in discrete, it may be presumed that its contribution as prerequisite is "maturity". (It is likely that, if students taking calculus after discrete were compared to other calculus students, those with discrete first would fare better.) One step has already been taken in addressing the problem of transfers: One Florida "common numbered" course previously accepted as prerequisite but rarely found to be sufficient will no longer be accepted. This will largely solve the "transfer gap" problem. The Florida State College Algebra and Trigonometry course contains some elementary symbolic logic and other topics which should be helpful. Both topical coverage and text are under examination. This important foundation course is rarely taught by regular faculty; the teaching assistants are being given additional help as they try to bring their students to the desired levels of mathematical maturity. The examination and reworking of this important "pre-calculus and pre-discrete" course will continue with a new text.

The Second Edition of the locally-written text for discrete (Mott-Kandel-Baker), which has been revised with help from the grant supporting this project, will be adopted for the academic year 1986–1987. The overall picture of student success and its implications will be reconsidered. The faculty teaching the discrete courses generally enjoy teaching them and are optimistic that improved backgrounds and new text materials will raise student achievement levels.

APPENDIX: Raw Test Items

Samples follow of final exams, hour tests and short quizzes, collected from eight instructors of Discrete I and II. Materials from both regular professors and from TAs and adjunct instructors are included. These are in "raw" form; they are a retyping of actual tests as administered and have not been edited or annotated. They are included to give an indication of the types of problems and the nature of the demands of the courses. These items should not be copied for tests or assignments without careful checking and, as necessary, addition of figures.

Discrete I Final Examinations

FSU SAMPLE DISCRETE I FINAL EXAM

1. Determine whether or not the pairs of graphs are isomorphic. If so, label the vertices of both graphs so that corresponding vertices have the same label. If not, explain fully why not. (Three pairs of graphs are given.)

2. The diagram shown is a subgraph of a binary tree. The level order index of the vertex w is 47. (i) Find the level order indices of the vertices labeled u and v. (ii) How many ancestors does v have in the whole binary tree (not just in the pictured subgraph)?

3. Determine if the following sequences are graphic. If so, draw a diagram of a graph with that degree sequence; if not, explain fully why not.

   (i) $(2,2,2,2,3,3)$

   (ii) $(1,3,3,3)$

   (iii) $(1,1,2,2,2,3,4,5,5,6)$

4. Suppose G is a graph all of whose vertices has degree 3. Suppose that $|E| = 2|V| - 3$. Find $|V|$.

5. Define a relation R on Z by xRy if $(x - y)/5 \in Z$, for $x, y \in Z$. Prove that R is an equivalence relation on Z.

6. Let G be a graph with n vertices and m edges such that all vertices have degree 6 or 7. Let N_6 be the number of vertices of degree 6 and N_7 the number of degree 7. Show that $N_6 = 7n - 2m$.

7. Let G be a connected graph with $|E| < |V|$. Prove that G is a tree.

8. A forest is a simple graph with no circuits.
   (i) Explain why any connected component of a forest is a tree.
   (ii) Suppose G is a forest that has 51 edges and 62 vertices. How many components does G have? Why?

9. Perform preorder, inorder, and postorder traversals on the binary tree pictured.

10. Tell me something interesting. Do not make any comment about the grades you hope to receive on this exam, the quiz, and the class.

FSU SAMPLE DISCRETE I FINAL EXAM

All solutions must be documented appropriately. For proofs, this means you must convince me that you understand. for computations, this means enough intermediate steps must be included to show how the answer is obtained.

1. Consider the inference $a \lor b$
   $$\neg a \lor c$$
   $$\neg c$$
   $$\text{therefore } \neg b$$
   a. The direct-proof test form is:
   b. The proof-by-contradiction test form is:
   c. Prove or disprove the inference given below using a truth table.

2. Suppose that the integers $1\ldots99$ (inclusive) are arranged in some unknown order around the circumference of a disk. Prove that some three consecutive numbers must sum to at least 150.

3. Find the number of integral solutions of $x+y+z = 8$ satisfying $0 \le x \le 4$, $0 \le y \le 3$, $0 \le z \le 2$.

4. a. Evaluate the sum $C(8,8)+C(9,8)+\cdots+C(20,8)$.
   b. Evaluate the sum $C(8,0)+C(9,1)+\cdots+C(28,20)$.

5. a. What is the smallest possible number of vertices of a graph with 7 edges if each vertex has degree at most 3?
   b. Why?

6. Decide which of the following pairs of graphs are isomorphic, and give reasons for your decisions. (Three pairs of graphs are given.)

7. A forest with 15 vertices and 9 edges has how many components?

8. a. Describe Kruskal's algorithm operating on a network G. Be sure to cover the three phases of the algorithm (initialization, loop, and termination); use vertex coloring as a control structure.
   b. What is the result of Kruskal's algorithm applied to an arbitrary network G?

9. a. Apply Prim's algorithm to the given network, beginning with vertex A using alphabetical order to make otherwise unspecified choices. Indicate the output of this algorithm by listing edges in order of discovery.
   b. What is the result of Prim's algorithm applied to an arbitrary network G?

10. What is the maximum height of a binary tree with n vertices?

11. What is the maximum height of a complete binary tree with n vertices?

FSU SAMPLE DISCRETE I FINAL EXAM

Caution: Use your time wisely and don't spend too much on the first, computational, portions. Watch the weightings. Give full explanations; carefully cite theorems used.

1. a. Is there a binary tree whose level order indices are 1,2,3,4,5,7,8,9,18,37? (If so, sketch; if not, why not?)
   b. Suppose T is a binary tree whose largest level order index is $4^{10} + 2$. What is the height of T?
   c. Suppose a vertex of a binary tree has level order index 213. What is the level order index of its parent? What is the level order index of its right child?

2. Suppose we have a graph G with the following edges and costs. Sketch G below; then find and sketch a minimal spanning tree; give cost.

Edge: $\{e,f\}\{e,c\}\{e,b\}\{d,c\}\{d,e\}$ $\{b,c\}\{a,c\}\{a,b\}$ $\{a,d\}\{f,b\}$
Cost: 2   6   4   3   7   10   4   7   1   5

3. Let G be a graph all of whose vertices have degree 4. Find $|V|$ and $|E|$ if $|E| = 3|V| - 8$.

4. How many integer solutions are there of $x_1 + x_2 + x_3 + x_4 + x_5 = 8$ where $x_1 = 1$, $x_2 = 1$, and x_3, x_4, x_5 are each greater than or equal 1?

5. How many arrangements are there of the letters of the word HIGHLIGHTS with
   a. no other restrictions?
   b. the G's adjacent?

6. Symbolically negate, using "$\forall$": $\exists x, [\neg F(x)]$:
   Symbolically negate, using "$\exists$": $\forall x, [F(x)]$:
   Give an equivalent symbolic statement using "$\forall$" to:
   $\neg[x, F(x)]$:

7. Enumerate the number of ways of placing 20 indistinguishable balls into 5 boxes where each box is nonempty.

8. You have gone with your family to Grandma's for Christmas. After the initial "welcome" supper, it is dull. In an attempt to brighten things up, you volunteer to set up decorations and begin by cutting an 11 foot tree on the back forty. Since the electric heater cannot be plugged into an extension cord and uses one of the two living room outlets, you have only 1 outlet. You are DESPERATE to connect Grandma's two living room lamps and the TV, the Joy-to-the-World doorchimes, Grandma's neon Nativity scene, the tree-top angel, 20 strings of lights and a corn popper. (No courtesy jacks. Plenty of fuses.) What is the least number of extension cords each having 4 outlets which will hook it all up? Draw the appropriate tree with extension cords as edges.

For #9-14, prove or disprove.

9. Suppose T_1 and T_2 are trees with no vertices in common. Let "a" be a vertex of T_1 and "b" be a vertex of T_2. Let T_3 be the graph consisting of all the vertices and edges of T_1 and of T_2 along with the extra edge $\{a, b\}$. Prove that T_3 is a tree.

10. An edge e in a connected graph G is a cut edge if $G - e$ is not connected. Prove that if a graph G is a tree, then each edge is a cut edge.

11. $0(n + \ln n) = 0(n)$.

12. The set of all subsets of a given set is called the power set. For the (sub)sets in the power set, a relation R is defined as ARB if $A \subseteq B$. R is an equivalence relation.

13. There are between $h - k$ and $(k^{h+1} - 1)/(k - 1)$ vertices in a directed tree of height h and degree k.

14. The graphs are isomorphic.

Of the proofs 15–18, do 3. You must indicate your omission.

15. A nondirected graph G is connected if G contains a spanning tree. (Hint: successive deletion of edges until no circuits remain.)

16. In a complete binary tree with n vertices, the indices of the vertices in the hth level comprise the completed interval 2^h through $2^{h+1} - 1$, or from 2^h through n if n is less than $2^{h+1} - 1$.

17. A tree with n vertices has exactly $n - 1$ edges.

18. Using the definition of 0, show that $2n + 11 \in 0(n)$.

Discrete I Hour Tests

Note: All instructors do not sequence topics in the same order. See "Project Summary," Case, paragraph 2.

FSU SAMPLE DISCRETE I TEST 1

1. Construct a truth table for the proposition $(p \vee q) \wedge [\neg(p \vee r)] \to q \wedge \neg r$ and decide whether or not this is a tautology.

2. Prove that the implication in (a) is a valid argument and produce a counterexample for (b) to show that the implication in (b) is invalid.
   (a) $p \to (r \to s)$
   $\quad \neg r \to p$
   $\quad p$
   $\quad \therefore s$
   (b) $p \to (r \to s)$
   $\quad \neg r \to \neg p$
   $\quad p$
   $\quad \therefore \neg s$

3. Use a contrapositive argument to prove that $a \cdot b \in D \to a, b \in D$ where a and b are integers.

4. Negate the following symbolic statement, passing the negation through the quantifiers and write your negated statement without the use of the negation symbol $\neg$. $\exists x, \forall \epsilon > 0, \exists t > 0, \forall s, s \geq t \to |f(s) - x| < \epsilon$

5. Negate each of the following sentences and write your final negated sentences in proper English.
   a. Some relations are not transitive.
   b. All prime numbers are not even.
   c. For every integer x, either $x^2/5 > 7$ or $x < 2$.

6. On a certain circular drive, there are 30 houses and 30 mailboxes. On a particular day, the mailman deposits at least 1 and no more than 30 letters in each mailbox. It turns out that no two mailboxes have the same number of letters in them. Show that there are 4 consecutive mailboxes that contain a total of at least 62 letters.

7. Prove that for each $n \in N$, $1 \cdot 1! + 1 \cdot 2! + \cdots + n \cdot n! = (n + 1)! - 1$.

8. Suppose that 88 chairs are arranged in a rectangular array of 8 rows and 11 columns. If 50 students are seated (1 per chair), prove that
   a. some row must have at least 7 students, and
   b. some column must have no more than 4 students.
   HINT: prove by contradiction!

9. Prove that $\forall n \in N$, 6 divides $n(n^2 + 5)$ evenly.

10. A runner has 30 days left to train for a marathon. She knows that she will require no more than 50 total hours of running time during those 30 days. She decides to run at least 1 hour each day and always to run an integer number of hours (for example, she will run for 3 hours or 4 hours, but not 3 1/2 hours). Prove that there is a sequence of consecutive days in which she has run a total of exactly 9 hours for those days.

FSU SAMPLE DISCRETE I TEST 2

1. How many integers between 1 and 107 contain exactly two 5's and one 9?

2. How many ways are there to select 3 cards in order from a deck of 52 if
   a. the first two cards are queens and the last is a spade?
   b. all cards chosen are face cards (J, Q, K) and the first card is a spade and the last two cards are not spades?

3. How many 5-card poker hands have
   a. a full house?
   b. a pair of aces and no other pairs?
   c. 3 of a kind (without another pair)?

4. How many numbers between 1 and 1008 have a sum of digits less than 7?

5. How many ways are there to place 20 balls in 8 numbered boxes so that exactly 3 boxes are empty and exactly one box contains exactly one ball (i.e., 3 empty boxes, 1 box with 1 ball, 4 boxes with at least 2 balls per box)?

6. A store sells shirts in 27 different colors. In how many ways can a customer buy 5 shirts if they are
   a. all of different color?
   b. not necessarily different colors?
   c. of exactly 2 different colors?

7. a. How many arrangements are there of the letters of the phrase TENNESSEE VOLUNTEERS?
   b. How many are there with no two E's adjacent?

8. Let n be a positive integer. Show by a combinatorial argument that

$$\binom{3n}{n} = 3\binom{n}{3} + 6n\binom{n}{2} + n^3.$$

   HINT: Consider n men, n women, and n children.

9. Use mathematical induction to prove the following special case of Diagonal Summation for all $n \geq 1$.

$$\binom{10}{0} + \binom{11}{1} + \cdots + \binom{10+n}{n} = \binom{11+n}{n}.$$

FSU SAMPLE DISCRETE I TEST 2

All solutions must be documented appropriately. For proofs, this means enough intermediate steps must be included to show how the answer is obtained.

1. Using the letters A, B, C, D, E how many words can be made if
   a. each letter must be used exactly once?
   b. no letter may be used more than once?
   c. the letters may be used any number of times but the word must contain exactly four letters, counting repeats?
   d. the word contains exactly 15 letters, and the letters may be used as follows: A-5 times, B-4 times, C-3 times, D-2 times, E-1 time?
   e. the word contains exactly 4 letters, and the letters may be used as in part (d) above?

2. Morse code is a language whose words are made using the two letters "$\cdot$" and "$-$" ("dot" and "dash"). How many codewords can be made using 5 dots and 3 dashes?

3. Find the number of integral solutions of:
   a. $x + y + z = 8 \quad 0 \leq x, 0 \leq y, 0 \leq z$
   b. $x + y + z = 8 \quad 3 \leq x, 0 \leq y, -5 \leq z$
   c. $x + y + z \leq 8 \quad 0 \leq x, 0 \leq y, 0 \leq z$
   d. $x + y + z = 8 \quad 0 \leq x \leq 4, 0 \leq y \leq 3, 0 \leq z \leq 2$

4. Give a combinatorial argument to show that $C(n-1, k-1) + C(n-1, k) = C(n, k)$.

5. a. Write $2 + 6 + \cdots + n(n+1)$ in terms of a column sum.
   b. Evaluate the sum $C(7,0) + C(8,1) + \cdots + C(27,20)$.
   c. Evaluate the sum $C(7,7) + C(8,7) + \cdots + C(20,7)$.

6. Find the coefficient of $w^6 x^3 y^7 z$ in the expansion of $(w^3 - x + 2y + z)^{13}$.

7. How many integers between 1 and 500 are not multiples of 5, 6, or 9?

FSU SAMPLE DISCRETE I TEST 1
Write your answers clearly and completely.

1. Prove one of the following by induction:
   a. $2^{n-1}(3^n + 5^n) > 8^n$ for all integers $n \geq 2$.
   b. 5 is a factor of $n^5 - n + 5$ for all integers $n \geq 2$.

2. Suppose that 13 mailboxes are arranged in a circle and 67 letters are distributed among them.
   a. Prove that one mailbox contains no more than 5 letters.

b. Prove that there is a pair of adjacent mailboxes which contain a total of at least 11 letters.

3. Disprove the following: For all integers n, if $n < 5$ then $n^2 < 25$.

4. Given 8 white, 7 blue, 6 red, 10 black and 4 yellow balls,
   a. how many balls must be chosen to guarantee there are at least 3 of some color chosen?
   b. how many balls must be chosen to guarantee there are at least 2 white, 5 blue, 3 red, 8 black and 1 yellow ball chosen?

5. Write the negation to the following by changing the quantifier.
   a. All math majors are either smart or not lazy.
   b. There exists an integer n such that $n^2 + 7$ is divisible by 3.

6. Write the contrapositive to the following: If it is sunny and warm, we will have a picnic.

7. Write the arguments in symbolic form and determine if they are valid. If so, state the rules of inference which justify your answer. If not, identify the fallacy.
   a. If the governor signs the bill $[S]$, he will retain the labor vote $[L]$. The governor retained the labor vote. Therefore, the governor signed the bill.
   b. If the governor vetoes the bill $[V]$, he will lose the farm vote $[F]$. He has retained the farm vote. Therefore, he did not veto the bill.
   c. If the governor signs the bill $[S]$, he will lose the farm vote $[F]$. He will lose the election $[E]$ if he loses the farm vote. He will not sign the bill. Therefore, he will lose the election.

8. Prove that given 7 positive integers, there exist two of them whose sum or else whose difference is divisible by 10. (HINT: place the integers in boxes labeled 0, 1, 2, 3, 4, 5 where an integer is placed in box r if it can be written in the form $10n + r$ or $10n + (10 - r)$ for some positive integer n.)

FSU SAMPLE DISCRETE I TEST 2

1. In how many ways can 6 girls and 5 boys
   a. sit in a row?
   b. sit in a circle?
   c. sit in a row if girl A and boy B must sit in adjacent seats?
   d. sit in a row if girl A and boy B must not sit in adjacent seats?
   e. sit in a circle if the girls must sit together?

2. How many 5-card hands from a deck of 52
   a. have 5 spades?
   b. have 5 cards in one suit?
   c. have exactly 3 aces and no other pair?
   d. have exactly 1 pair?

3. How many integral solutions are there of $x_1 + x_2 + x_3 + x_4 = 20$ where $x_1 \geq 2$, $x_2 \geq -1$, $x_3 \geq 5$, $x_4 \geq 0$?

4. From a group of 5 physicists and 4 mathematicians how many ways can
   a. a committee of 5 be formed?
   b. a committee of 2 physicists and 3 mathematicians be formed?
   c. a committee of 4 be chosen if it must include at least one physicist and at least one mathematician?
   d. the group be divided into teams where the first team has 3 members, the second team has 4 members and the third team has 2 members?
   e. the group be divided into 3 teams with three members each?

5. How many ways are there to arrange the letters in the word BOOKKEEPER
   a. with no other restrictions?
   b. with the E's adjacent?
   c. with no two of the E's adjacent?

6. A shop sells 10 different flavors of ice cream. In how many ways can a customer choose 5 ice cream cones (one dip of ice cream per cone) if they
   a. are all different flavors?
   b. are not necessarily of different flavors?
   c. contain exactly 3 different flavors?

7. Give a combinatorial argument to prove the following:
   a. $(n - 3) \cdot C(n - 3) = n \cdot C(n - 1, 3)$ $(n \geq 4)$
   b. $(6n)!/(3![(2n)!]^3)$ is an integer

FSU SAMPLE DISCRETE I TEST 3

1. Define the relation R on the set of integers by xRy iff xy is even. For each property listed, write "yes" if R has the property or write "no" and give a counterexample.
   a. Transitive
   b. Reflexive
   c. Symmetric

2. Prove that $7x + 5x^2$ is in $0(2x^2)$.

3. Determine if the following sequences are graphic. If so, draw a simple graph with such a degree sequence. If not, explain why not.
   a. $(1, 1, 2, 2, 5)$

b. $(2, 2, 2, 4, 4)$

c. $(2, 2, 3, 3, 3)$

4. A forest F with 3 components has 12 edges. How many vertices does F have?

5. Let G be the graph:
   a. List the cut edge(s) of G
   b. Draw the subgraph induced by the vertices of degree 3
   c. Draw the complement G of G

6. Given the digraph G below with the subgraphs G_1 and G_2, answer the following questions about each of G, G_1 and G_2.
   a. Is it weakly connected?
   b. Is it quasi-strongly connected?
   c. Does it have a directed spanning tree?

7. Are the following pairs of graphs isomorphic? If so, give a correspondence between the vertices describing the isomorphism. If not, explain why an isomorphism is impossible. (Two pairs of graphs are given.)

8. Five desert outposts must be connected by water lines. The connection costs between each pair of outposts is given in the chart (in thousands of dollars). Find the cheapest way to construct the water line.

	a	b	c	d	e
a	0	1	5	3	2
b		0	7	1	4
c			0	5	4
d				0	2
e					0

9. Prove or disprove the following statements. Clearly indicate which three you choose.
   a. If G is a graph with $|E(G)| = |V(G)| - 1$, then G is a tree.
   b. If two non-adjacent vertices of a tree T are connected by an edge, then the resulting graph will contain a circuit.
   c. If G is connected and has n vertices, then G has at least $n - 1$ edges.

Discrete I Quizzes

FSU SAMPLE DISCRETE I QUIZ 2
Prove the following by the method indicated.

1. If x and y are numbers such that $3x + 9y = 61$, then either x or y is not an integer. (Direct proof)

2. Suppose that 63 seats in a classroom are arranged in 7 parallel rows of 9 seats each (so that there are 9 columns with 7 seats each). Suppose that 40 students are seated randomly in the classroom. Show that some row has at least 6 students. (Prove by contradiction.)

FSU SAMPLE DISCRETE I QUIZ 4
Indicate how you computed your answer.

1. How many 4-digit numbers can be formed using the digits 3, 4, 5, 6, 7 if
   a. repetitions of digits are allowed
   b. repetition of digits is not allowed
   c. repetitions are not allowed and the number is divisible by 5

2. How many ways are there to select 2 cards without replacement from a deck of 52 cards if
   a. the first card is an ace and the second a queen
   b. neither the first nor the second card is an ace
   c. at least one of the cards is an ace
   d. at least one of the cards is not an ace

3. List all 3-combinations and 2-permutations of $\{0 \cdot a, 1 \cdot b, 2 \cdot c\}$.

FSU SAMPLE DISCRETE I QUIZ 5
Circle your answer—it may involve factorial notation. Show work.

1. In how many ways can 4 boys and 3 girls sit in a row
   a. if the girls must sit together?
   b. if the boys and girls are to have alternate seats?
   c. if the children in the end positions must have the same sex?

2. How many ways can a committee of 3 teachers and 2 students be chosen from a group of 7 teachers and 5 students?

3. Find the number of integral solutions to $x_1 + x_2 + x_3 + x_4 = 50$ where $x_1 \geq -4, x_2 \geq 7, x_3 \geq -14, x_4 \geq 10$.

4. How many ways are there to place 20 identical balls into 6 different boxes in which exactly 2 boxes are empty?

5. How many ways are there to arrange the letters in the word "Canadian"?

FSU SAMPLE DISCRETE I QUIZ 6
Show your work.

1. In how many ways can 5 boys and 6 girls sit in a row if
   a. the boys must sit together
   b. the children in the end positions must not have the same sex?

2. How many ways can a committee of 2 teachers and 5 students be chosen from a group of 6 teachers and 11 students?

3. Find the number of integral solutions to $x_1 + x_2 + x_3 = 20$ if $x_1 \geq -1$, $x_2 \geq 4$, and $x_3 \geq 2$.

FSU SAMPLE DISCRETE I QUIZ 10

1. Given the graph G as indicated, draw the following:
   a. G', the complement of G
   b. the subgroup induced by the vertices of G of degree 3

2. Determine whether the following pairs of graphs are isomorphic. If so, give the isomorphism. If not, explain why not. (Two pairs of graphs are given.)

FSU SAMPLE DISCRETE I OPEN HOMEWORK QUIZ

1. Complete the statement of the major theorem we proved for today which begins "If x and m are relatively prime positive integers, then   "

2. Write out the equivalence classes resulting from application of the above theorem for $w = 3$, $x = 23$, $m = 6$. The theorem assures us that these equivalence classes are___. As a matter of fact, a 1–1 correspondence exists (because they are =!) between these classes and the members of ___. Write out this equality.

3. If possible, solve for x below. If not, explain why not.
   a. $[x] \cdot [7] = [2]$ in Z_6
   b. $[4] \cdot [x] = [2]$ "mod 8"

FSU SAMPLE DISCRETE I QUIZ (4.4.1)

Look closely at graphs G_1 and G_2. Answer for each. Is it a partial ordering? If not, why not? Is it a total ordering? If not, why not? Is it a well ordering? If not, why not? If so, indicate the order of the elts (and hence the minimal elt).

Discrete II Final Examinations

Two final exams are given. Also of interest is the first set of four hour tests in the following section. These were given during a summer term and each after the first contains material from earlier units(s); there was no final examination for that section.

FSU SAMPLE DISCRETE II FINAL EXAM

1. a. The characteristic polynomial for $a_n - 11a_{n-1} + 39a_{n-2} - 45a_{n-3} = 0$ is ___.
   b. One characteristic root is $t = 5$. List all 3 characteristic roots.
   c. List the form for $a_n = $ ___.
   d. Find the unique solution when $a_0 = -4$, $a_1 = 4$, $a_2 = 20$.

e. List the form of a particular solution to: $a_n - 11a_{n-1} + 39a_{n-2} - 45a_{n-3} = f(n)$ where
$f(n) = 3$
$f(n) = (n+5)3^n$
$f(n) = (n+3)2^n$
$f(n) = (n+3)5^n$

2. Solve the divide-and-conquer relation $a_n - 5a_{n/3} = n$ when $a_1 = 2$. Use "change of variables".

3. A signaling device can send 7 types of signals. Types 1 and 2 take 1 clock cycle to transmit; types 3, 4, 5 take 3 clock cycles to transmit; types 6, 7 take 5 clock cycles to transmit. Find a recurrence relation for the number an of signals which can be sent in n clock cycles.

4. Give definitions for:
   a. $x \equiv y \bmod 53$
   b. A relation R on a set A is an equivalence relation iff
   c. An $S - O$ cut $(X, \overline{X})$ is a minimal cut iff
   d. A path P in a transport network is flow-augmenting for a flow F iff

5. a. Use the Euclidean algorithm to find an integer x, $0 \leq x < 53$, where $39x \equiv 1 \bmod 53$.
   b. Find an integer y where $0 \leq y < 12$ such that $7266 \equiv y \bmod 13$

6. Let $A = \{a, b, c\}$ where $a < b < c$. Order the strings $a, b, c, ab, ac, aab, aa, abc, bc, aac, bba$ using
   a. lexicographic ordering
   b. enumeration ordering

7. List all possible topological sortings of the vertices of the following. (A digraph with loops is given.)

8. Compute the adjacency matrix for the transitive closure of the relation R whose adjacency matrix is

$$M(R) = \begin{bmatrix} 0 & 0 & 1 & 0 & 0 \\ 0 & 0 & 0 & 0 & 0 \\ 0 & 1 & 0 & 0 & 1 \\ 0 & 0 & 1 & 0 & 0 \\ 1 & 0 & 0 & 0 & 0 \end{bmatrix}$$

(Use Warshall's algorithm and list each intermediate matrix.)

9. i. For the given flow F, find $|F|$
   ii. For $X = \{S, a\}$, find $F(X, \overline{X}), k(X, \overline{X})$
   iii. Indicate a maximal flow G on the graph and list the value of this maximal flow G, $|G|$

iv. List all minimal cuts and explain your reasoning (A network flow is given.)

10. Let $A = \{2, 3, 4, 6, 8, 12, 24, 36\}$. Let R be the relation defined by $(a, b) \in R$ iff a divides b; i.e., there is an integer c such that $b = ac$.

   a. List the poset diagram for R (also called the Hasse diagram).

   b. List the maximal elements under R.

   c. List the minimal elements under R.

11. Show that if A is a set containing n elements, then there are 2 to the power $n^2 - n$ reflexive relations R on A. (HINT: consider the adjacency matrix for R.)

12. If R is the relation defined on the set of rational numbers by $(a, b) \in R$ iff $a - b$ is an integer, prove that R is transitive.

FSU SAMPLE DISCRETE II FINAL EXAM

1. Find the coefficient of x^{27} in $(x^2 + x^3 + x^4 + x^5 + x^6)^{10}$.

2. Suppose $a_0 = 2$, $a_1 = 11$, and $a_n = 5a_{n-1} - 6a_{n-2}$ for $n \geq 2$. Find an explicit formula for a_n: $a_n = \underline{\quad}$. Check you answer by computing a_2 using the formula and using the recursion.

3. Find the form of the particular solution if $a_n - 4a_{n-1} + 4a_{n-2} = (5n^2 + 1)2^n$.

4. A signaling device can send 7 types of signals. Types 1 and 2 take 1 clock cycle to transmit; types 3, 4, 5, 6, take 3 clock cycles to transmit; type 7 takes 4 clock cycles to transmit. Find a recurrence relation for the number an of signals which can be sent in n clock cycles.

5. a. Define: $x \equiv y \bmod 79$

   b. Use the Euclidean algorithm to find an integer x, $0 \leq x \leq 79$, with $68x \equiv 1 \bmod 79$.

6. Let $A = \{x, y, z\}$ with $x < y < z$. Order $y, x, xz, yx, xxy, xy, xx$ from smallest to largest, using enumeration ordering.

7. Let $V = \{a, b, c\}$ with $a < b < c$. Let $R = \{(a, b), (b, b), (c, b)\}$, $S = \{(a, b), (b, c), (c, a), (c, c)\}$

   (i) Find and arrange in lexicographical order:

   a. The join of $R\&S$ along component 1 of $R\&$ component 2 of S

   b. $S \circ R$

   (ii) Find the adjacency matrices A and B of R, S respectively, and compute A or B.

   (iii) In general, if A and B are adjacency matrices of relations R, S on an ordered set V, then A or B

is the adjacency matrix of___

   A and B is the adjacency matrix of___

   A or and B is the adjacency matrix of___

8. Find *all possible* topological enumerations of the vertices. (Digraph with loops is given.)

9. i) $|F| = \underline{\quad}$

   ii) For $X = \{S, a\}$, $F(x, \bar{x}) = \underline{\quad}$
       $k(x, \bar{x}) = \underline{\quad}$

   iii) The value of a maximal flow is___ (A network flow is given.)

10.(i) State Hall's Marriage Theorem.

   (ii) Use network methods to find a maximal matching. If the maximal matching is not a complete matching, use the labelling algorithm to show that the hypothesis of Hall's Marriage Theorem does not hold.

Discrete II Hour Tests

The first four tests were given to a summer section; the last three are cumulative in lieu of a final exam.

FSU SAMPLE DISCRETE II TEST 1

1. a. How many equivalence relations are there on a set $\{a, b, c, d, e, f\}$ of 6 distinct vertices, such that there are at least 3 vertices that are equivalent to each other? (This part of the question is not asking for non-isomorphic equivalence relations.)

   b. Among the equivalence relations counted in part (a), how many non-isomorphic ones are there?

2. Of the 6 special properties of relations we studied, which ones does the following relation have, and which ones does it not have?

   xRy if and only if x and y are integers and either $x \leq y$ or $x = 2y$

3.
Given the 5 functions at the right, construct the 5×5 matrix A whose i,jth element is 1 if $f_i \in 0(f_j)$ and is 0 if $f_i \in 0(f_j)$.

$f_1(x) = 1000$
$f_2(x) = x^2$
$f_3(x) = x \log_{10} x$
$f_4(x) = x$
$f_5(x) = (1.5)^x$

4. Let $P = \{(x, 3x-2) | x$ an integer$\}$ and $Q = \{(x, x^2) | x$ an integer$\}$. For each of the following relations, list all elements of the relation if it has fewer than 6 elements; otherwise list 6 of its elements. [For example, if one of the questions concerned P, which has infinitely many elements, one might list the 6 elements $(5, 13)$, $(10, 28)$, $(20, 58)$, $(0, -2)$, $(100, 298)$, $(-5, -17)$.]

a. $P \circ Q$, the composition of P and Q

b. the join of P and the relation $R = \{(1, 7), (0, 4), (5, 11)\}$, with respect to the second element of each of the relations

5. For the digraph at right, construct the adjacency matrix M or.and M.

6. Apply Warshall's algorithm to the adjacency matrix A of the digraph at right, showing the matrix as it appears after each major step.

FSU SAMPLE DISCRETE II TEST 2

1. Is the graph at right planar? Prove that your answer is correct. (Do not merely show a graph, but give a step by step description of what you do, and why.)

2. Let G be a polyhedral graph such that every region of G has degree ≥ 5, and such that there are exactly 53 regions. Prove that the number of vertices must be ≥ 82.

3. Apply Grinberg's Theorem to the graph at right. What can you conclude from this theorem for this graph?

4. Does the graph at right have a Hamiltonian circuit that includes edge c_i? List carefully the steps you follow, giving a complete argument.

5. For the graph in problem 4,
   a. Is there an Euler circuit? Why?
   b. Find the chromatic number, giving a careful argument.

6. a. Find the coefficient of $x^6 y^9$ in the expansion of $(x^2 + 5y)^{12}$.
   b. In the expansion of $(x_1 + x_2 + x_3 + x_4 + x_5)^{20}$, there are ___ terms, and the coefficient of $x_1^3 x_2^2 x_3 x_4^7 x_5^7$ is ___.

7. Draw all the subgraphs with 4 vertices and 3 edges, up to isomorphism, of the graph at the right. (That is, do not draw 2 subgraphs that are isomorphic.) How many of these non-isomorphic subgraphs are there?

FSU SAMPLE DISCRETE II TEST 3

1. In a certain school of 100 students, 40 are taking French, 40 are taking German, and 40 are taking Latin. 10 are taking both French and Latin (these may or may not be taking German). 20 students take only French, while 20 take only Latin, and 15 take only German. How many are taking all three languages? How many take no languages?

2. How many 13-card bridge hands have at least one ace, at least one king, at least one queen, at least one jack, and at least one ten? (Each of the 4 suits has 13 cards: 2, 3, 4, 5, 6, 7, 8, 9, 10, Jack, Queen, King, Ace.) You must use *Inclusion-Exclusion*.

3. Find a generating function for a_r, where a_r is:
   a. the number of solutions of $e_1 + e_2 + e_3 + e_4 = r$ such that each e_i is an integer greater than 0; e_2 and e_3 are odd, and e_4 is ≥ 3.
   b. the number of ways to select r sandwiches from among 3 kinds of sandwich, if we have an unlimited number of sandwiches of the first kind but only 100 of each of the other two kinds, and must choose at least 3 sandwiches of the third kind.

4. Find the coefficient of x^{12} in $(1 + x)^5/(1 - x)^5$.

5. Find the coefficient of x^5 in the expansion of $(3 + x)/(1 - 7x + 10x^2)$.

6. Find a recurrence relation for a_n, the number of n-digit ternary sequences having no pair of consecutive 0's. (Ternary sequences use only 0's, 1's and 2's as digits; a ternary sequence need not use all three of these digits.)

7. Use Warshall's algorithm to compute the adjacency matrix of the transitive closure of the digraph given. Show the matrix as it appears initially and as it appears after each major step.

8. a. Using Euler's formula, give a proof that a polyhedral graph cannot have exactly 7 edges.
   b. Either carry out a similar proof for 11 edges (instead of 7), or show where the proof fails in the case of 11 edges.

FSU SAMPLES DISCRETE II TEST 4

1. Find the number of integers in the set $\{1, 2, 3 \ldots, 1000\}$ that are divisible by at least one of 4, 5, or 6. You are required to use Inclusion-Exclusion.

2. No matter how one draws a Hamiltonian circuit for the graph at right, there will always be exactly the same number x of the small squares inside the circuit. Find this number x, and use Grinberg's Theorem to show why there will always be exactly this number x of small squares inside the circuit.

3. Solve $a_n - 3a_{n-1} - 4a_{n-2} = 0$, $a_0 = 1$, $a_1 = 1$, by means of a generating function. (You must use the method of generating functions.)

4. Using the method of characteristic roots (not using a generating function),

a. give the general solution of $a_n - 8a_{n-1} + 16a_{n-2} = 0$

b. the solution of $a_n - 8a_{n-1} - 65a_{n-2} = 0$ such that $a_0 = 0$ and $a_1 = 2$.

c. the general solution of a homogeneous linear recurrence relation with constant coefficients, whose characteristic roots are 1, 1, 1, −2, −2, 3, 4, 4, 4, 4.

5. For a certain inhomogeneous linear recurrence relation with constant coefficients, the roots of the characteristic polynomial (of the related homogeneous equation) are 1, 1, 2, 3, 3, 3. If the inhomogeneous term $f(n)$ is as given in each of the following parts, give the *form* of a particular solution a_n^p of the inhomogeneous equation.

a. $f(n) = 7\,5^n$

b. $f(n) = (7n^2 + 3n)5^n$

c. $f(n) = 5$

d. $f(n) = n^4$

e. $f(n) = (n^2 + 6)3^n$

6. Solve $a_n - 5a_{n-1} + 6a_{n-2} = (-1)^n$, $a_0 = 0$, $a_1 = 1$.

FSU SAMPLE DISCRETE II TEST 1

1.

i. Find the generating function

$$A(x) = \sum_{r=0}^{\infty} a_r x^r$$

where a_r is the number of ways of putting r identical balls into 12 bins, if no bin is empty, the even-numbered bins contain an even number of balls each, and bin 6 has no more than 8 balls. (Do not simplify your answer.)

ii. Find the coefficient of x^{23} in

$$(x^2 + x^3 + x^4 + x^5 + x^6 + x^7)^8(1 + x + x^2 + x^3 + \cdots)^9$$

2. Solve $a_n + a_{n-1} = 3n$, $a_0 = 5$ using the method of undetermined coefficients.

3. i. Find the form of the particular solution of $a_n - 5a_{n-1} + 6a_{n-2} = (5n^2 + 1)2^n$

ii. The form of a particular solution of $a_n - 5a_{n-1} + 6a_{n-2} = 3^n$ is $a_n(p) = An3^n$. Find the particular solution.

4. Express the generating function for $a_n + 2a_{n-1} + 10a_{n-2} = 0$, $a_0 = 7$ $a_1 = 16$ in the form $A(x) = (a + bx)/(c + dx + ex^2)$. (That is, carry out the solution to the point where partial fractions would be required.)

5. Find a recurrence relation for the number an of valid strings of length n using the symbols A, B, 1, 2, 3 if no more than 2 consecutive symbols may be numbers. (Thus, $ABA12$ is valid, but $1A213B$ is not). Compute: $a_1 = \underline{\quad}$, $a_2 = \underline{\quad}$, $a_3 = \underline{\quad}$, $a_4 = \underline{\quad}$

FSU SAMPLE DISCRETE II TEST 2

1. Let A be the set of positive integers. Define a relation R on A by: $(m, n) \in R$ iff there exists an integer i (positive, negative, or zero) with $m/n = 2^i$. Show that R is an equivalence relation on A.

2. Let A be the set (a, b, c) with $a < b < c$. List aaa, a, ab, ba, baa, bab, aba, cba, cb from smallest to largest according to
   i. lexicographic ordering
   ii. enumeration ordering

3. i. Let $m > 1$ be a positive integer. Define: $x \equiv y \bmod m$, for any integers x and y.
   ii. Show that $-x \equiv m - x \bmod m$, for any integer x
   iii. Use the Euclidean Algorithm to find an integer x, $0 \le x < 73$, with $66x \equiv 1 \bmod 73$ (HINT: use (ii) at the end)

4. Let $A = \{a, b, c\}$ with $a < b < c$. Let $R = (a, c)$, (c, a), (b, c), $(b, b)\}$, $S = \{(b, a), (b, c), (c, b)\}$.
   i. Find the join of R and S along component 1 of R and component 2 of S.
   ii. Find $R \circ S$
   iii. Find the adjacency matrix A of R. Then compute A or and A. Then compute A or $(A$ or $\circ$ and $A)$. Write (in terms of R and operations on relations) the relations whose adjacency matrix is A or $(A$ or $\circ$ and $A)$.

5. i. State Warshall's Algorithm for computing the transitive closure R^+ of a relation R.
   ii. Employ Warshall's algorithm for the relation R of problem 3, listing the matrices M_0, M_1, M_2, M_3 and using M_3 to write R^+ as a set of ordered pairs.

6. Using the definitions $x \oplus y = \max(x, y)$

$$x \otimes y = \begin{cases} \text{if } x > 0, y > 0 \\ 0 \text{ otherwise} \end{cases}$$

Compute $\begin{matrix} 9 & 2 & 1 \\ 3 & 4 & 0 \\ 5 & 6 & 1 \end{matrix} \oplus \cdot \otimes \begin{matrix} 0 & 5 & 2 \\ 6 & 7 & 0 \\ 1 & 12 & 1 \end{matrix}$

FSU SAMPLE DISCRETE II TEST III

1. Give the sequence of values for i, $l(i)$, $u(i)$, $m(i)$ computed using the Binary Search algorithm with input $x = 29$ and $A = \langle 1, 3, 7, 10, 13, 14, 18, 20, 22, 26, 30, 47\rangle$

2. State the Interchange Sort algorithm. Using this algorithm, the number of comparisons required to sort a list of 100 items is about ___. Implement the algorithm to sort the list $\langle 3, 1, 5, 4, 2\rangle$, describing each step.

3. Implement the Topological Sort algorithm to get a topological enumeration of the vertices of the digraph with loops below.

4. Use the augmenting flow chain method to find a maximal flow. List all augmenting flow chains that you use, showing slack along each edge and the slack along the entire chain. The value of the maximal flow is ___. Find a cut (x, x) of minimal capacity.

$$\text{edge} \quad \text{cap.} \quad \text{flow}$$

$(x, \bar{x})$
$(\bar{x}, x)$

$k(x, \bar{x}) = \underline{\quad}$
$F(x, \bar{x}) = \underline{\quad}$
$F(\bar{x}, x) = \underline{\quad}$

5. (i) State Hall's Marriage Theorem.
   (ii) Use network methods to find a maximal matching, and (if such exists) a subset C of A so that $R(C)$ has fewer elements than C does.

Discrete II Quizzes

FSU SAMPLE DISCRETE II QUIZ 2
Write down the generating function for each problem.

1. Put identical balls into 4 boxes with at most 4 balls in box 1, at least 2 balls in box 2, any number in boxes 3 and 4.

2. Put identical balls into 10 boxes with 2, 3 or 6 balls in box 1, the number of balls in box 2 a multiple of 5, the remaining boxes non-empty.

3. Put balls which may be red, blue, or green into 10 boxes. No more than 2 balls in any box.

FSU SAMPLE DISCRETE II QUIZ 3
Let a_n be the number of ways to tile a strip of length n, using red and white tiles of length 2, and blue, yellow and green tiles of length 1. (Not all colors need be used.) Find a recurrence relation for a_n, and compute:

$$a_1 = \underline{\quad}, a_2 = \underline{\quad}, a_3 = \underline{\quad}, a_4 = \underline{\quad}$$

FSU SAMPLE DISCRETE II QUIZ 3
Solve using the method of undetermined coefficients:

$$a_n + a_{n-1} - 30a_{n-2} = 0, a_0 = 1, a_1 = 27$$

FSU SAMPLE DISCRETE II QUIZ 5

Use the Euclidean Algorithm to find an integer x with $227x \equiv 1 \mod 1000$.

FSU SAMPLE DISCRETE II QUIZ 6
Let $A = \{\vee, \wedge, \neg, \rightarrow\}$ with order $\vee < \wedge < \rightarrow < \neg$

Arrange $\wedge \rightarrow$, $\vee\vee \rightarrow$, $\neg \vee \vee\wedge$, $\rightarrow$, $\neg\vee \rightarrow$, $\vee\neg\vee$, $\vee\neg \vee \neg$ from the smallest to largest using
   i) lexicographic ordering
   ii) enumeration ordering

FSU SAMPLE DISCRETE II QUIZ 7
Let R_1 and R_2 be the relations on $A = \{a, b, c\}$ given by $R_1 = \{(a, a), (a, b), (b, b), (c, b)\}$ $R_2 = \{(a, a), (a, c), (b, c)\}$
   Find
   (i) $R_1 \circ R_2$
   (ii) $R_2 \circ R_1$
   (iii) R_2^{100}

FSU SAMPLE DISCRETE II QUIZ 8

i) Compute the adjacency matrix A of the graph given.

ii) Define $x \oplus y = \max(x, y)$

$$x \otimes y = \begin{cases} x + y & \text{if } x > 0, y > 0 \\ 0 & \text{otherwise} \end{cases}$$

$$\text{Compute} \quad \begin{matrix} 0 & 9 & 1 \\ 4 & 0 & 5 \\ 0 & 3 & 2 \end{matrix} \oplus \cdot \otimes \begin{matrix} 4 & 2 & 8 \\ 0 & 0 & 6 \\ 5 & 7 & 0 \end{matrix}$$

FSU SAMPLE DISCRETE II QUIZ 9

1. Give the sequence of values for i, $l(i)$, $u(i)$, $m(i)$ computed using the Binary Search algorithm with input $x = 13$ and
$A = \langle 1, 3, 7, 10, 13, 14, 18, 20, 22, 26, 30, 47\rangle$

2. Implement the Merge Sort algorithm, beginning with the sorted subsets $\langle 2, 5\rangle$, $\langle 1, 3, 5, 7\rangle$, $\langle 4, 6, 10\rangle$, $\langle 2, 3\rangle$, $\langle 0, 7, 9\rangle$.

FSU SAMPLE DISCRETE II QUIZ 10
Implement the Topological Sort algorithm to extend the relation $R = \{(a, c), (a, d), (b, a), (c, d), (d, d), (e, a), (e, b), (e, e)\}$ to a total order on the set $A = \{a, b, c, d, e\}$.

FSU SAMPLE DISCRETE II QUIZ 11
Use the labelling algorithm to augment the given flow to a maximal flow F. List any augmenting flow path that you find, showing the slack on each and the slack along the path. Find a cut $(X, \overline{X})$ of minimal capacity (using the algorithm), and calculate (showing your work) $|F|$, $F(X, \overline{X})$, $k(X, \overline{X})$, and $F(X, \overline{X})$

Montclair State College

Prepared by Kenneth Kalmanson

Review of the 1983-1984 School Year

The first year of our project at Montclair State College went entirely according to plan. We randomly selected one half of the available pool of eighty mathematics and computer science majors for our project. (The other half were assigned to calculus 1, the traditional sequence being calculus 1,2,3, after which computer science majors, the majority, would then take a semester of discrete mathematics, while mathematics majors would take a semester of linear algebra.)

The project group was divided into two sections. Dr. William Parzynski taught one section and Dr. Kenneth Kalmanson (the author of this report and project director) taught the other. Both instructors are full time faculty members of the Department of Mathematics and Computer Science at Montclair State College in New Jersey. We each used the textual materials in discrete mathematics prepared by Dr. Kalmanson. Topics covered during the first semester included properties of sets, functions and numbers; mathematical induction and other proof techniques; mathematical logic, logic gates, and logical circuits; combinatorics and recursion; properties of graphs and digraphs, including trees; and some topics in network analysis. Algorithms were discussed in connection with the majority of topics and some corresponding computer software was developed and demonstrated.

The attrition rate in these freshman discrete mathematics sections was about the same as we usually get in freshman calculus. About 75% of the combined sections finished the course with a grade of "C" or better. The percentage was a bit lower in my own section, perhaps because I was using the same materials concurrently in a discrete math section composed of sophomores and juniors. The performance level and sophistication of my freshmen was somewhat lower than that of the upperclassmen, in spite of the fact that no calculus was required in either class. I attribute this difference, at least in part, to the fact that some marginal students who might have otherwise found their way into my discrete math section for upperclassmen had been winnowed out in freshman calculus. Moreover, I suspect that a year or more of calculus (and other college level courses) can only have had a positive effect on the students' study habits, mental discipline, and mathematical sophistication.

The second semester of our project also went according to plan. The linear algebra course presented in both sections of our pilot program included systems of equations, matrices, determinants, the vector spaces R^2 and R^3 in some detail, followed by general vector spaces, dot and cross products, and linear transformations. We began by applying matrices to graphs, and we concluded with applications to Markov chains and intuitive ideas about limits of sequences. Once again, computer software was developed and used to demonstrate algorithms for matrix operations, applications, and the like.

I should mention that our programs have been written in Basic because we could not assume that our students had prior knowledge of programming or a more algorithm-oriented language such as Pascal at the beginning of the course. In fact, our students were all in the process of learning PL-1, but we wanted to avoid mixing up problems of computer formatting and syntax with those of formulating an algorithm. At any rate, most students seemed to have little difficulty following Basic programs - and we did not require any programming from them at all.

The performance level of our freshmen in the linear algebra component as measured by grades was, once again, pretty good, with about 80% achieving a grade of "C" or better. I myself, however, did not give any grades of "A". Moreover, some topics in our traditional linear algebra course, such as inner product spaces and diagonalization, had to be omitted for lack of time and student sophistication.

It should be noted at this point that many of the mathematics majors who completed a year in the program decided to switch to the usual three semester calculus sequence rather take than the modified two semester sequence we had planned for them. We felt it only fair to offer them this choice since they would (as mathematics majors) be required to take three semesters of calculus in either case. According to our original plan, however, computer science majors in the program were required to take only two semesters of calculus. This caused some unhappiness among most of the mathematics majors. Still another difficulty was the fact that many of the program's participating students had already had some calculus in high school. For these students, having to delay taking college calculus for a year often sparked feelings of resentment. We decided, therefore, to try to alleviate these difficulties by the way we would select the next incoming freshman class, as we shall describe below.

Review of the 1984-1985 School Year

For the reasons that we have just discussed, the second year's freshmen population differed from the first in both size and composition. We began the year 1984-1985 school year with one discrete math section of 27 freshman computer science majors and no mathematics majors, of whom three dropped out before the first semester ended, as compared to the two sections of about 40 mathematics and computer science majors of the first year.

Moreover, rather than take a random sample of one half of the eligible freshmen, we consciously selected only those freshmen who had no background in high school calculus.

As expected, the morale of the 1984-85 freshman class was much higher than the previous year's class. The lack of calculus does not seem to have affected the grades of the 1984 discrete math class, which were 4 A's, 8 B's, 4 C's, 5 D's, and 3 F's. Although the 1985 linear algebra grades were also similar - with 2 A's, 5 B's, 7 C's, 3 D's, and 3 F's - to those given in the 1984 class, it must be admitted that we were not able to cover as much material in either class as in the previous year. Furthermore, we have not been able to cover as much material in these freshman sections of discrete math and linear algebra as with classes composed largely of sophomores, falling short of proposed syllabi (see attached) by a few sections in each course. (Had we moved at the pace we maintain in sophomore level courses, there would certainly have been lower grades.)

Of the original 44 students selected for our pilot in 1983, only 9 entered our calculus 1 class in September of 1984. The attrition was due to the failing of one of the previous two courses, or to students dropping out of school or out of their major, or deciding that they would rather not continue in the pilot program even though they continued as math or computer science majors.

The students who remained in our program for a second year were highly motivated and, as one might have expected, did rather well in terms of grades. We should add that the entire revised syllabus for both calculus 1 and calculus 2 was covered, including material on sequences, series, and multivariate calculus. There is significantly more material in our revised calculus 2 course than in what has been the standard curriculum at Montclair State. Therefore, the good grades these students earned should be considered something of an

achievement. (4 A's, 3 B's, 1 C, and one withdrawal in calculus 1; and 3 A's, 3 B's, and 1 C in calculus 2.)

An attempt was made to determine if the discrete mathematics and linear algebra courses had had any effect on the students' performance in their Foundations of Computer Programming courses. An analysis of the means and variances of our program's students' grades versus other students in the programming courses seemed to show a small significant difference in favor of our program's students. This is especially interesting in light of the fact that there was no attempt to integrate the programming courses with our pilot program's courses and the heterogeneous nature of the programming courses's populations. However, the difference between the two groups was so small and there were so many confounding factors that one should not, we feel, make much of this. Moreover, the very different backgrounds of the pilot population and the regular population the second year preclude a similar analysis of freshmen grades in 1984-85.

It might be more beneficial to study the effects that other factors may have had on determining success in our discrete math and linear algebra courses. In particular, we would like to determine if there is a correlation between verbal scores on the SAT's and the grades students earn in these courses. We suspect that reading deficiencies indicated by low verbal scores tend to hinder a student in discrete math even more than, say, in calculus.

Continuation of the Project

At the conclusion of the 1984-1985 school year I recommended that the department of mathematics continue our pilot program for another year, and, indeed, two sections of freshman discrete math were scheduled to be offered in the fall of 1985, as well as a section of our revised calculus sequence for sophomores. However, a sharp drop in our incoming computer science freshmen class forced us, during the summer of 1985, to reassess and either offer our new curriculum to all students or to none. For reasons described below, we have decided to offer only the traditional curriculum to freshmen in the fall of 1985. The sophomore calculus 1 section, however, is running as planned.

We will now list what we perceive to have been the major strengths and weaknesses of our program at Montclair State College

Weaknesses of the Program

1. We have already mentioned the lack of integration with the Foundations of Computer Science (program-

ming) course. We had to assume that our students had no programming knowledge because they were incoming freshmen. Our presentation of software written in Basic compensated for this a bit, but it might have been better to have had the students begin the pilot program after they had had one term of $PL - 1$, the language taught in Foundations of Computer Science 1.

2. Students had continuing difficulty in writing algorithms.

3. A high attrition rate left us with very few students in the pilot's Calculus 1 and 2 the first year. We have a somewhat larger Calculus 1 class in the fall 1985 semester (sixteen students), even though we began with a smaller freshman population in 1984. The quality of the 1985 class appears to be down, as this report is being written.

4. The students seemed to find the linear algebra even harder than the discrete math, especially when we discussed more abstract notions such as vector spaces and linear transformations, as opposed to calculational problems such as, for example, Gauss-Jordan elimination. Consequently, we covered less material in the pilot course than in the standard course.

We have, over the years, discussed the desirability of having a strictly calculational linear algebra course at the freshman or sophomore level which would be followed by a more abstract vector space - linear transformation course at a higher level. But as we do not have such courses at present, we have been forced to find an acceptable compromise for our sophomore level course. However, I do not feel that the pilot's linear algebra course would be an acceptable compromise, since we do not yet have a second, more advanced linear algebra course available for the students. A better solution might be achieved by means of a single expanded linear algebra course which would incorporate both the intuitive foundation of two-and-three dimensional vector geometry as well as applications of eigenvalues.

5. The discrete math and linear algebra courses did not give the students nearly as much opportunity to use their high school geometry, algebra, and, especially, trigonometry as does the calculus sequence. Moreover, while the delay in using these skills may not hurt the very good student, it does seem to hurt the average student.

6. The high school mathematics curriculum has calculus as a natural goal. Many students are thrown off balance when they discover that they are not yet ready for calculus, and this can create classroom difficulties.

7. Students who have gotten credit for one semester of calculus before entering college had the additional problem with our program of either having to interrupt their study of calculus or staying out of the program entirely, that is, beginning their study of discrete math in the sophomore year. This would continue to be a problem if the pilot had been adopted as the standard program.

Strengths of the Program

1. There may be some positive transfer of learning between the discrete math and the Foundations of Computer Science courses, taken as a whole. The mean grade in Foundations 1 and 2 was 2.69 (out of 4) for the Sloan group, as opposed to 2.19 for the non-Sloan group. An analysis of the variances barely indicated significance (at a 0.05 level of confidence), but the sizes of the populations (23 vs. 34) were a little too small to attach much meaning to this. Furthermore, there were some glaring instances of students doing well (A or B) in discrete math and getting D or F in the Foundations courses.

2. All of the courses given in the pilot project used computer-assisted demonstrations to a much greater extent than in comparable non-pilot courses. This seems to have been well received by the students, but whether these demonstrations added much to their knowledge, as opposed to their entertainment, is questionable. Unfortunately, there were insufficiently many personal computers available for us to assign computer-assisted homework assignments, which I believe would have made the classroom demonstrations of software more valuable.

3. Computer science students in our program took linear algebra. This course was not required at the time for computer science majors.

Benefits to the MSC Mathematics Program

1. The one-year calculus sequence, partly as a result of our successful experience this year, will now be standard for computer science majors, beginning with the spring 1986 semester. This one year sequence will include infinite series, but it will not include an introduction to multivariate calculus, which computer science students can take as an elective in calculus 3.

2. For our own purposes at Montclair State College, we have demonstrated the desirability of beginning the students' college mathematics study with calculus, rather than with discrete mathematics, given the present national academic environment.

3. Computer science majors will also be required to take linear algebra and probability (as of 9/86), another development stimulated in part by our pilot project. (They are already required to take discrete mathematics.) Moreover, we are now in the process of changing our sophomore level linear algebra course so that it follows the recommendations given above: One more semester hour; more geometric intuition; and more advanced applications.

4. The discrete mathematics text developed for our pilot was published by Addison-Wesley in 1986 under the title *An Introduction to Discrete Mathematics and Its Applications*. The material was generally well-received by both the students in our pilot and others who have used it in the traditional course. The computer science programs will be made available free of charge by Addison-Wesley to those who adopt the textbook.

Conclusions

Since our pilot program has not been adopted as the standard curriculum at Montclair State College, it is tempting to dismiss it as a failure. However, the insights obtained by running our pilot program have enabled us to make several improvements in our standard program that might not have been otherwise made. Moreover, we have achieved our major goal of correcting the balance between the number of required analysis courses and discrete mathematics courses where we deem it appropriate, namely, for the computer science majors. Indeed, I think it fair to say that the major mathematics offerings in the first two years at Montclair State College have been considerably strengthened as a result of our learning from both the weaknesses and the strengths of our Sloan Foundation supported pilot. We are grateful to the Sloan Foundation for their support and encouragement during this time of transition.

Syllabus: Calculus I (MSC Sloan Fnd Project)

The following syllabus assumes that the student has taken four years of high school mathematics, including intermediate algebra, trigonometry, logarithms, and exponents.

# of Lectures	Topic
	I Functions and Graphs
1	Coordinates and Distance
1	Circles and the Slope of a Line
1	The Equation of a Straight Line
1	Graphs of Monomial and Quadratic Functions
	II The Derivative
1	The Slope of a Graph
2	Limits
2	Differentiation of Polynomials
2	More Limits and Continuity
	III Differentiation and Differentials
2	Differentiation of Products and Quotients
2	The Differential
1	The Chain Rule
1	Higher Order Derivatives
2	Applications to Motion
2	Implicit Differentiation
	IV Trigonometric Functions
1	Review : Evaluation and Identities
1	Review : Graphs
1	Limits
1	Derivatives
	V Applications of Derivatives
2	Related Rate Problems
1	Newton's Method
1	Relative Maxima and Minima
1	Absolute Maxima and Minima
1	Mean Value Theorem
2	Signs of Derivatives and Curve Sketching
1	Limits at Infinity and Asymptotes
2	Extremal "Word Problem"
2	Antiderivatives and Substitution
	VI The Definite Integral
1	Review of Summation Notation
1	Areas and Approximations
2	Riemann Sums and the Definite Integral
1	The Fundamental Theorem of Calculus
1	Integration and Simple Differential Equations
1	Integration Using Tables

#		#	
	VII <u>Applications of the Definite Integral</u>	2	Applications to Moments and Centroids
1	The Area Between Two Curves	1	Trapezoidal and Simpson's Rule
1	Averages and the Mean Value of a Function	1	Improper Integrals
		2	Indeterminate Forms

Total number of lectures = 47

The plan given above assumes a 50 minute lecture. It leaves ample time for at least four 50 minute exams and a two-hour final, in addition to one review lecture for each of the seven major topics. The instructor may choose to use some of the review time for computer demonstrations using prepared software from Conduit ("Arbplot") and Addison-Wesley (by Finney, et. al.). This would be particularly appropriate in connection with graphing, limits, Newton's method, and Riemann sums.

The text used for the above syllabus was John B. Fraleigh's "Calculus with Analytic Geometry" (1st ed), Addition-Wesley.

Syllabus: Calculus II
(MSC Sloan Fnd Project)

The following syllabus assumes that the student has completed a Calculus I course that included, in particular, trigonometric functions,the Fundamental Theorem of Calculus, and applications of the definite integral to areas between two curves and to the average value of a function.

# of Lectures	Topic
	I <u>Applications of Integration</u>
1	Volumes of Revolution: Washers
1	Volumes of Revolution: Shells
1	Arc Length
	II <u>Other Elementary Functions</u>
2	Natural Logarithm Function
2	Exponential Function
2	Other Bases and Logarithmic Differentiation
2	Applications to Growth and Decay
2	Inverse Trigonometric Functions
2	Hyperbolic Functions and Their Inverses
	III <u>Techniques of Integration</u>
1	Integration by Parts
2	Partial Fractions
2	Integration Tables: Reduction Formulas and Completing the Square
	IV <u>Further Topics in Integration</u>

#	Topic
	V <u>Infinite Series of Constants</u>
1	Convergence of Sequences
1	Convergence of Series
1	Integral Test
1	Comparison Tests
1	Ratio Test
1	Alternating Series
1	Absolute Convergence
	VI <u>Power Series</u>
1	Definition of Power Series
1	Taylor's Formula
2	Taylor's Series
	VII <u>Topics in Geometry</u>
2	Parametric Equations: Curves and Derivatives
2	Polar Coordinates: Curves and Derivatives
1	Areas in Polar Coordinates
1	Coordinates in Space
	VIII <u>Topics in Multivariate Calculus</u>
2	Partial Derivatives
1	Maxima and Minima
1	Integrals over a Rectangle
1	Integrals over a Region

Total = 48

The plan given above assumes a 50 minute lecture. It leaves ample time for at least four 50 minute exams and a two-hour final, in addition to at least one review session for each of the 8 major topics. The instructor may choose to use some of the "review" time for computer demonstrations. Given the inevitable days lost due to holidays not made up, "snow days", and just getting bogged down from time to time, one should not expect to have time left over from the 60 class meetings allocated to this course. We suggest using John B. Fraleigh's *Calculus with Analytic Geometry* (Addition-Wesley, 1980) as a text.

Syllabus: Linear Algebra
(MSC Sloan Foundation Program)

# of Lectures	Topic
	I <u>Systems of Linear Equations and Matrices</u>
1	Introduction to systems of linear

	equations
2	Gaussian elimination
1	Homogeneous systems of equations
1	Review of matrix operations
1	Matrix algebra
1	Elementary matrices
1	Finding A^{-1}
1	Further results on A^{-1} and consistency

II Determinants

1	Determinants, permutations, and elementary products
1	Evaluation by row reductions
1	Properties of determinants
1	Cofactor expansion
1	Cramer's rule

III Vectors in 2-Space and 3-Space

1	Geometric introduction
1	Coordinates and translations
1	Distance and norms
1	Dot product; projections
1	Cross product
2	Lines and planes in space

IV Vector Spaces

1	Euclidean n-space
2	General vector spaces: axioms and examples
1	Subspaces
1	Linear independence
1	Basis and dimension
1	Row and column spaces of a matrix
1	Change of basis

V Linear Transformations

1	Definitions and examples
1	Properties of linear transformations
1	Kernel and range
1	Geometry of linear transforms from R^2 to R^2
1	Matrices of linear transforms
1	Similarity

VI Applications and a Transition to Calculus

2	Markov processes and matrices
1	Intuitive introduction to limits
1	Limits and Markov chains

Total = 39

The outline given above assumes 45 lectures (or class meetings) in a three credit/hour course, with 50 minutes allowed for each lecture. It allows three lectures for review and three lecture sessions for exams. The organization of the course follows Howard Anton's Linear Algebra (3rd and 4th ed.). It assumes that the student will take calculus I as the next course.

Syllabus: Discrete Math. Structures (MSC Sloan Foundation Program)

Neither calculus nor computer programming experience are required for this course. However, it is assumed that the student has had at least four years of high school mathematics in order to have developed the necessary level of skill in handling verbal problems and algebraic manipulation, as well as general mathematical sophistication.

# of Lectures	Topic

I Sets, Numbers, and Algorithms

1	Defining sets and subsets
2	Sets and functions
2	Sums and algorithms
1	Integers and algorithms
1	Rational and real numbers
1	Other systems of numeration
1	Binary arithmetic and twos complements

II Sets, Logic, and Computer Arithmetic

2	Examples, counterexamples, and math. induction
1	Set operations and their applications
2	The algebra of set operations
1	Truth sets and truth tables
2	Laws of logic and rules of reasoning
1	Logic gates and computer arithmetic

III Applied Modern Algebra

2	Boolean algebras
2	Minimization and switching circuits

IV Combinatorics and Recursion

1	The multiplication principle and permutations
2	Combinations and binomial coefficients
2	Recurrence models
1	Closed form solutions by backwards substitution
1	Analysis of algorithms

V Introduction to Graph Theory

2	Graphs, Euler trails, Hamiltonian paths

1	Classification of graphs	
1	Graph isomorphisms	
1	Matrices and graphs	
2	Matrix multiplication and	
	connectedness	

VI Trees and Digraphs

1	Weighted graphs, spanning trees and the connector problem
1	The shortest path problem
1	Directed graphs

Total = 39

The outline given above assumes 45 lectures (or class meetings) in a three credit/hour course, with 50 minutes allowed for each lecture. It allows three lectures for review and three lecture sessions for exams. The organization of the course follows Kenneth Kalmanson's course notes (published in 1986 as *An Introduction to Discrete Mathematics and Its Applications*, Addison-Wesley.) Software has been developed to illustrate various parts of the course, but students are not expected to write programs.

Math 122 - Calculus   Dr.Kalmanson

Final Examination

Do all work in your answer book. You have two hours for this examination.

I. For each of the following, find the limit, if possible (as a number or plus or minus infinity), or state that the limit does not exist.

(a) $\lim_{h \to 0} \frac{(3+h)^2 - 9}{h}$

(b) $\lim_{x \to 2+} f(x)$ if

$$f(x) = \begin{cases} x^2 - x & \text{if } x < 2 \\ 3 & \text{if } x = 2 \\ x - x^2 & \text{if } x > 2 \end{cases}$$

(c) $\lim_{x \to \infty} \frac{3x^2 + x - 1}{x - x^3 + 10}$

(d) $\lim_{x \to 0} \frac{3x}{\sin x}$

(e) $\lim_{x \to 0} \cot x$

II. Is the function in I(b) continuous? Explain briefly.

III. Evaluate each of the following:

(a) $D_x(\cos^2 4x)$ (b) $D_x[(\sec x)(\tan x)]$

(c) $f'(0)$ if $f(x) = (1 + x + x^2 + x^3)(x^4 + x^5 + x^6 + x^7)$.

(d) $f'(0)$ if $f(x) = (x + 1)/(x - 1)$.

(e) $D_x^2(\sin x)$.

(f) The slope of the tangent line to the curve $x^2 y + y^2 - x = 1$ at the point $(x, y) = (0, 1)$.

(g) The equation of the normal line to the graph of $y = x^2$ where $x = 3$.

IV.(a) Graph the function $y = x + [1/(x + 1)]$, showing all relative minima, relative maxima, inflection points, and asymptotes, if any.

(b) Find the maximum value of $y = x + [(1/(x + 1)]$ on the interval $[-5, -3/2]$.

V. Do any four of the following problems:

(a) Find the average value of the function $y = \tan x$ on the interval $[\pi/6, \pi/4]$.

(b) Use Newton's method and three iterations to estimate a zero of the function $f(x) = x^2 + 3x + 1$.

(c) Find the maximum area that a rectangle can have if its perimeter is 20ft.

(d) If the radius of a circular disk is increasing at a rate of 3 inches/second, find the rate of increase in the area of the disk when its radius is 10 inches.

(e) The height of a particle at time t seconds is given by $f(t) = -16t^2 + 100$. Find the velocity of the particle when the particle is at ground level and falling.

VI. Do any four of the following:
   (a) State the Mean Value Theorem for derivatives.
   (b) State the Mean Value Theorem for integrals.
   (c) Prove that if a function $f(x)$ is differentiable at $x = x_0$, then it is continuous at x_0.
   (d) Prove that if a function $f(x)$ is continuous on $[a, b]$ and differentiable on (a, b), such that $f'(x) > 0$ on (a, b), then $f(x)$ is increasing on (a, b).
   (e) Prove that, if $f'(x) = 0$ throughout $[a, b]$, then $f(x) = a$ constant on $[a, b]$.

VII. Find the area of the region between the curves $y = \sin x$ and $y = \cos x$ from $x = 0$ to $x = \pi$.

Math 221(s) - Calculus II   Dr. Kalmanson

Final Examination

Answer each question in your answer booklet. You have two hours in which to complete this examination.

I. Find the (a) radius and (b) interval of convergence of the power series

$$\sum_{n=1}^{\infty} \frac{(x - 2)^n}{n3^n}$$

II.(a) Write the Taylor polynomial of degree 7 for $f(x) = \sin x$ in powers of $x - \pi$.
   (b) Show that the remainder $|E_7(\pi + \pi/180)| < 1/100,000$.

III. Write the Taylor's series expansion of four of the following functions about $x = 0$:
   (a) $\sin x$ (b) $\sinh x$
   (c) e^{-x} (d) $x/(1 - x)^2$
   (e) $ln(1 - x)$ (e) $\tan^{-1} x$

IV. Let $f(x, y) = x^2 + y^3 + xy$.
   (a) Find f_{xy}.
   (b) Write equations of a line tangent to the surface $z = f(x, y)$ at the point $(x, y, z) = (2, 1, 7)$ in the plane $y = 1$.

V. Find the volume of the solid obtained by revolving the region bounded by the curve $y = 1/x$, the line $x = 1$, and the x-axis about the x-axis.

VI. Do two of the following:
   (a) Find the area inside the intersections of the regions bounded by the curves $r = \sin \theta$ and $r = \cos \theta$.
   (b) A thin sheet of material has a constant mass density of 3 units and just covers the first quadrant region bounded by the curves $y = x^2$, $y = 1$, and $x = 0$. Find its first moment about the y-axis
   (c) Find the length of the arc given by the parametric equations $x = 2 \cos t$ and $y = 2 \sin t$ for $\pi/3 \le t \le \pi/2$ by means of integration.
   (d) The velocity at time t of a body traveling on a line is given by $v = \sin((\pi/2)t)$ ft./sec. Find the total distance the body travels from time $t = 0$ to time $t = 4$.

VII. Evaluate four of the following:
   (a) $\int (e^x - e^{-x})/(e^x + e^{-x}) \, dx$ (b) $\lim_{x \to -\infty} (\tan^{-1} x)$
   (c) $d/dx(\sin^{-1}(x^2))$ (d) $\lim_{x \to \infty} (1/x) \ln(x)$
   (e) $\lim_{n \to \infty} \sum_{k=0}^{n} 3(0.1)^{-k}$ (f) $\int_0^{\pi} |\cos x| \, dx$

VIII. Give a careful definition of one of the following:
   (a) The sequence $\{a_n\}$ converges to L.

(b) The series $\sum_{n=1}^{\infty} a_n$ converges to S.

IX. Find an antiderivative of one of the following:
   (a) $e^x \cos x$ (b) $1/x(x^2 + 1)$

X. Discuss the convergence or divergence of one of the following series:
   (a) $\sum_{n=1}^{\infty} n/\sqrt{(n^3 + n)}$
   (b) $\sum_{n=2}^{\infty} 1/(n \ln n)$

Discrete Mathematics (0701-285)   Dr. Kalmanson

Final Examination

1. Complete each of the following:
   (a) A trail is defined to be a walk in a multigraph in which all of the $-----$ are distinct.
   (b) A complete graph on 20 vertices has exactly $-----$ edges.
   (c) The minimum number of edges that a connected graph on 20 edges can have is $-----$.
   (d) Every tree on 20 vertices has chromatic number $-----$.
   (e) If a connected planar map has 20 vertices and 15 regions, then it must have $----$ edges.
   (f) If K_n has an Euler circuit, then n must be $-----$.

2. Consider the set $S = \{a_n : a_n = a_{n-1} - a_{n-2}$ if $n = 2, 3, 4, \ldots$ and $a_0 = a_1 = 1\}$.
   (a) Is S finite or infinite? Explain briefly.
   (b) Is S discrete or not? Explain briefly.
   (c) Find a_{10}
   (d) Find the power set of $\{-1, 0, 1\}$.

3.(a) Evaluate $\sum_{k=0}^{100} (3) (2^k)$ if $2^{99} \approx (6.3)(10^{29})$.
   (b) Evaluate $p(x) = 5x^4 + 3x^3 - x^2 - x + 6$ in nested form for $x = -2$.
   (c) Write an algorithm that will test if an integer $n > 1$ is a prime.

4. Let $S(n)$ be the statement $\sum_{k=1}^{n} k^2 = n(n+1)(2n+1)/6$. Prove that $S(n)$ is true for all positive integers n by mathematical induction.

5.(a) Write the decimal number 43.125 in binary notation.
   (b) Write the binary number $(11011.101)_2$ as a decimal number.
   (c) Write the binary number in part b as a hexadecimal number.
   (d) Write the decimal number 183.375 in scientific notation to four significant digits.
   (e) Write the two's complement of the binary number 111101 with respect to 2^6.

6. Let A, B, and C be subsets of U such that $|A| = 24$, $|B| = 30$, $|C| = 52$, $|A \cap B| = 12$, $|A \cap C| = 18$, $|B \cap C| = 20$, and $|A \cap B \cap C| = 8$.
   (a) Draw a Venn diagram and indicate $\overline{A} \cap (B \cup \overline{C})$.
   (b) Find $|\overline{A} \cap \overline{B} \cap \overline{C}|$.

7.(a) Prove that the statement $[(p \to q) \wedge p] \to q$ is a tautology.
   (b) State in symbols the principle of "modus tollens."
   (c) State one of DeMorgan's laws for propositions.

8.(a) Define a Boolean algebra in terms of its operations.
   (b) Give a Karnaugh map for the following Boolean expression $(x + y)'z + xy + yz'$.

9.(a) Expand the binomial $(x^2 - 2y)^5$ by the binomial theorem.
   (b) How many subsets of size 3 does the set $\{c, o, m, p, u, t, e, r\}$ have?

(c) How many four letter sequences can be made from the letters of the word "computer" without repetition?

(d) Write a recursive relation and initial conditions for the number of ways an that a person can climb n steps if he can climb 1, 2, or 3 steps up at a time.

10. Give an example, using exactly 5 vertices and at least 4 edges of each of the following:

(a) an Eulerian graph that is not Hamiltonian.

(b) a Hamiltonian graph that is not Eulerian.

(c) a non-planar graph.

(d) a complete bipartite graph.

(e) a planar map that does not satisfy Euler's formula.

11.(a) Find an Euler trail in the graph given below:

(b) Find a minimum spanning tree in a weighted graph having the distance matrix given below:

(a)

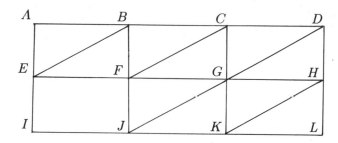

(b)

	A	B	C	D	E	F	G	H
A	–	4	6	14	16	15	20	12
B	4	–	5	14	13	14	16	11
C	6	5	–	17	16	15	18	10
D	14	14	17	–	9	8	7	14
E	16	13	16	9	–	1	2	18
F	15	14	15	8	1	–	3	15
G	20	16	18	7	2	3	–	17
H	12	11	10	14	18	15	17	–

Final Examination

Do each of the following problems in your answer book. You have two hours for this examination.

I. (a) Use elementary row operations on the augmented matrix to find all solutions of the following system of equations:

$$x + y + z + w = 0$$

$$x - y + z - w = 0$$

$$2x + 2z = 0.$$

(b) What is the dimension of the solution space in part (a)?

(c) If a linear transformation $T : R^4 \to R^3$ is defined by

$$T(v) = \begin{bmatrix} 1 & 1 & 1 & 1 \\ 1 & -1 & 1 & -1 \\ 2 & 0 & 2 & 0 \end{bmatrix} v,$$

what is the range of T, i.e. $T(R^4)$?

II. Use Cramer's Rule to find y in the unique solution of the system of equations

$$x + y - z = 3$$

$$x + 2y + z = 5$$

$$2x - y + z = 1$$

III. Prove: If x_1 and x_2 are solutions of the system of equations $Ax = B$, then $x_1 - x_2$ is a solution of the associated homogeneous system of equations.

IV. Prove: If U and W are subspaces of a vector space V, then $U \cap W$ is also a subspace of V.

V.(a) Find an equation of the plane containing the points $(1, 1, -2)$, $(1, 0, 1)$, and $(0, 1, 0)$.

(b) Find parametric equations of the line determined by the points $(1,1,-2)$ and $(1,0,1)$.

VI. If $u = (1, 2, 4)$ and $v = (-1, 3, 0)$, do each of the following:

(a) Find $u \cdot v$

(b) Find $u \times v$

(c) Is u orthogonal to v? Explain.

VII. Which of the following sets of vectors forms a basis for the vector space R^3? Explain in each case.

(a) $(1, 1, 0)$, $(0, 0, 1)$, $(-1, -1, -1)$.

(b) $(1, 1, 0)$, $(0, 0, 1)$

(c) $(0, 1, 0)$, $(1, 3, 0)$, $(1, 0, 3)$, $(1, 5, 0)$.

VIII.(a) Find a transition matrix P which will take one from the standard basis B for R^3 to the basis $B' = \{(0, -1, 0),\ (0, 0, -1)\ (1, 0, 0)\}$.

(b) Show that B' is an orthonormal basis.

(c) Find P^{-1}.

IX. Use the Gram-Schmidt procedure to obtain an orthonormal basis for R^2 from the basis $\{(1, 2),\ (3, 4)\}$.

St. Olaf College

Prepared by Loren Larson, Kay Smith, and Paul Zorn

Highlights

The main accomplishment of the St. Olaf Sloan-supported project in discrete mathematics in 1984–85 was the development of a freshman-sophomore curriculum which gives our students a more balanced exposure to discrete and continuous mathematics. At the core of this reform is a new sophomore-level course in discrete mathematics. In addition, we have endeavored to introduce a discrete viewpoint in our other introductory courses. We have also developed a new course, Applied Calculus, which emphasizes numerical methods and use of computer symbolic manipulation.

The new discrete mathematics course, together with the two-semester freshman calculus sequence and linear algebra, introduces students to the broad range of more advanced topics. At the same time, these courses provide a solid foundation for advanced work. This new beginning curriculum has broad support from our faculty and has been well received by students.

Discrete Mathematics

The level, content and goals of the discrete mathematics course that we developed are determined in part by the nature of St. Olaf College and our mathematics program. Because St. Olaf is a liberal arts college, most students take a wide variety of courses in the first two years. To encourage interaction in the classroom, class sizes are relatively small. The student body is well prepared and capable; most freshmen are ready for calculus. Mathematics is a popular subject at St. Olaf. For the last few years, more than ten percent of the graduating class have been mathematics majors. Many students who are not majors take two or three mathematics courses. There are no majors in computer science, engineering, or business. Students may complete a computer science concentration in conjunction with any major at the college. Most students do not take computer science courses in the freshman year.

When the Sloan grant was awarded, most potential majors and others with a serious interest in mathematics began with a year of single-variable calculus followed by a semester of linear algebra. In the second semester of the sophomore year, students took one or more of multivariable calculus, elementary real analysis, and computer organization. Courses in operations research and combinatorics were (and still are) offered as electives in the junior and senior year.

Discussions preceding the preparation of the Sloan grant proposal showed that department members were reluctant to make any changes in the curriculum which would reduce the interest in or flexibility of the mathematics program at St. Olaf. Some felt that adding more introductory courses might decrease enrollments in advanced courses and possibly reduce the depth of the major. Others felt strongly that calculus should remain the entry point for students who wanted to study mathematics and who did not have advanced placement. Faculty members like the coherence of the linear algebra course and consequently did not want a year-long course integrating discrete mathematics with linear algebra. It was also agreed that sophomore courses should assume greater mathematical maturity than freshman courses.

Starting with these departmental positions, we have developed a one semester sophomore-level discrete mathematics course. The current syllabus, described in the course outline, has evolved as we have considered what topics to include and tried different ways of presenting and organizing the topics. The course has a prerequisite of one semester of calculus and may be taken either before or after linear algebra. Because of the emphasis we place on problem-solving and proof, we view discrete mathematics as a course of benefit to all mathematics students, rather than as a course directed specifically at computer science students.

The course has three major goals:

i) to broaden students' knowledge of the range of mathematics by introducing them to some topics in discrete mathematics,

ii) to lay the foundation for upper-level courses by giving explicit attention to problem-solving methods and proof techniques, and

iii) to increase students' understanding of the mathematical process by encouraging them to formulate and test conjectures and by introducing open questions.

The topics covered were chosen because of their inherent interest, their importance in upper-level mathematics courses and their ability to illustrate problem-solving methods and proof techniques. We chose to focus on three major areas—modular arithmetic, counting, and graphs and trees—rather than to provide glimpses of many different topics. We devote sections to induction, recursion, searching for patterns, direct and indirect proof, backtracking and algorithms, and then use these ideas in the sequel. Some algorithm design strategies, such as greediness and divide and conquer, are described. We briefly discuss analysis and verification of algorithms. (See the course outline and comments for more details on content and rationale.) We chose not to cover algebraic structures since our students study abstract vector spaces in linear algebra, and all majors take abstract algebra.

The sophomore discrete mathematics course is accepted by the faculty as valuable for many students. Discrete mathematics seems to be a very good course for students who enter our program with advanced placement. It is a requirement for the computer science concentration and for secondary teaching certification. At present, faculty do not want to require the course for the mathematics major. Most do not consider it to be as fundamental as calculus and linear algebra. Perhaps this view will change as more faculty teach the course and as discrete mathematics itself becomes more central to mathematics.

Student response to the course has been generally favorable. Of students who completed course evaluation forms, nearly all responded that they would recommend the course to others. Asked how the course affected their attitudes toward mathematics, several students described the course as "fun." Others said that they liked being exposed to new areas different from calculus. Comparing the discrete mathematics course with calculus, a majority found the former more interesting, and all but a few found it at least as interesting. A majority rated discrete mathematics as more difficult than calculus. Surprisingly, half of the respondents rated discrete mathematics as more useful than calculus, while only a few considered calculus more useful. We had included some applications in the course but certainly did not emphasize its utility.

From the instructors' point of view, some aspects of the course are troublesome. In contrast to a calculus course, in which one begins with the limit, develops the derivative and integral as limits, and relates the concepts through the Fundamental Theorem, the discrete mathematics course lacks ongoing development of concepts and major theorems. We have endeavored to use problem-solving methods as a unifying theme by introducing the major ideas of induction, recursion, and algorithm early in the term and using these methods throughout the course. We have also included problems that require combining concepts from different parts of the course, e.g. counting and graphs.

A second problem is how to treat algorithms. Some of the students taking the course have no programming experience, while others have taken one or two computer science courses. Students familiar with programming wanted the algorithms as detailed as possible, while the novices were often baffled by pseudocode presentations of algorithms. Pseudocode using Pascal-like structures reveals more clearly the logical structure, particularly of recursive algorithms, but one cannot present full details in many cases without discussing data structures. We also found it difficult to devise problems related to the algorithms, other than straightforward applications of the algorithms to specific cases.

The Calculus Project

The main objective of the Sloan project at St. Olaf has been to expose students beginning their mathematical careers to discrete as well as continuous ideas and techniques. Although we considered diverting time from calculus to discrete mathematics (perhaps by condensing three semesters of calculus into two), we decided against doing so, for several reasons:

- Our calculus course is already relatively compressed. It meets three hours each week for thirteen weeks of instruction per semester.

- Because our multivariable course is both relatively advanced and also relatively detached from the first two semesters of calculus, it competes less directly with the discrete mathematics course than it would otherwise. Most mathematics majors delay multivariable calculus at least one semester after freshman calculus. (Sophomores usually take linear algebra rather than calculus in their first semester; many go on to elementary real analysis, discrete mathematics, and other courses before returning to multivariable calculus.)

- First-semester calculus has a natural structure, both as a major course and as a service course for other users. Therefore, reducing calculus from three semesters into two really means condensing the last two semesters into one.

- We agree that calculus as it is has problems, but we do not think they can be solved by shortening or deemphasizing the course.

Some of the problems (and possible solutions) we see with calculus have to do with the discrete-continuous debate. For example, calculus, as it is, focuses too narrowly on closed-form methods applied to explicit algebraic functions. More attention to discrete viewpoints

—approximation, error estimation, numerical integration, etc.—would broaden calculus' scope and display more of its power. Calculus also suffers from an undue emphasis on techniques. Students spend too much time learning to perform algorithms (e.g., formal differentiation, convergence testing) and too little understanding the theory underlying the formal techniques. A shorter calculus course might run even more to recipes than what we have now, because pressure to "cover" the standard canon of topics would remain.

A better alternative than shortening calculus is to strengthen the connection between theory and technique—each reinforcing the other. One way to do this is to use the theory of calculus to analyze the standard techniques (i.e., algorithms) that arise in calculus. Comparing numerical integration algorithms, for example, is an appealing application of calculus and a chance to introduce the big oh formalism.

Calculus has a special relationship with discrete mathematics. The theory of calculus is largely the interplay of discrete and continuous ideas. Moreover, students going anywhere beyond the most elementary discrete mathematics need calculus. What they need from calculus is not the standard compendium of closed-form techniques applied to explicit algebraic functions but a real mastery of calculus ideas (limit, rate of change, growth of functions, etc.) From the opposite point of view, the discrete viewpoint can help extend calculus to include more of its most interesting ideas and applications.

Given these views, our main effort in the calculus part of the Sloan project has been to include more "discrete" viewpoints (broadly defined) in calculus, without reducing its traditional theoretical content. The main tangible product so far is a collection of annotated problem sets. Most of them concern a "discrete" aspect of some idea in calculus (limit of sequence vs. limit of function, polynomial approximations to transcendental functions, numerical solutions to equations, etc.) Some of them extend a standard idea or result (e.g., the chain rule) beyond the usual elementary-function setting to graphical, tabular, and other "inexplicit" functions.

Implicit in the problem sets is the belief—borne out by experience! – that radical changes in mainstream calculus courses do not occur quickly. Therefore, the problem sets are supposed to be self-contained enough to work as supplements to almost any standard calculus course, and thus perhaps to aid in a gradual evolution of calculus in the directions outlined above. Calculus courses are already "full;" adding more material is simply not possible. Moreover, we have not found large "obsolete" sections of calculus to excise. Choices

need to be made: is another day on numerical integration worth omitting center-of-gravity calculations? The problem sets aim to give instructors such choices.

To complement the long-term, evolutionary development of mainstream calculus, we developed a new course—Applied Calculus—intended mainly for students between their first two semesters of calculus. The new course aims to augment traditional "exact" calculus with discrete and numerical ideas. Two unusual features of the course are its emphasis on approximate, discrete, and numerical methods in calculus, and heavy use of the computer symbolic manipulation package SMP. Many of the calculus problem sets described above are used as part of the course material. Students also used a calculus-level SMP user's guide, written at St. Olaf.

One reason to use symbolic computation in calculus is that studying numerical and approximation methods in calculus requires a great deal of formal and numerical manipulation. The standard Simpson's rule error estimate, for example, involves an extremum of the fourth derivative of the integrand. For most functions one would wish to integrate numerically, the mechanics of finding the extremum are both too difficult for a freshman and to some degree beside the point. With SMP, on the other hand, the student can easily (and correctly!) compute the higher derivatives, graph them, and, if necessary, solve equations numerically. Then the original error estimation problem can be addressed for its own sake. Another, perhaps less obvious, advantage of computer algebra systems is that they handle numerical calculations without the distraction of programming. Computing numerical approximations to integrals by hand, for example, is clearly impractical. Basic and Fortran programs can help, but they require distracting programming effort. With programs like SMP, students need to learn only the syntax (not the programming) of such commands as SIMPSON $[f[x], x, 0, 1, 10]$ to compute an approximation to an integral over $[0, 1]$ with 10 subdivisions. (Students easily learn and use such simple commands even in a mainstream calculus course that makes no deep or regular use of symbolic computation.)

The Applied Calculus course was well received, judging by the student evaluations. A substantial majority rated the course "very good" or "excellent", and nearly all respondents said they would use SMP in further courses. All respondents felt that the course helped them understand calculus ideas better.

The Sloan project at St. Olaf has not yet radically altered the calculus experience for the average student. The new Applied Calculus course, for example, is not required of majors—both time and computer capacity are obstacles. What we have done is to rethink quite carefully what we want from calculus and why, especially in light of new interest in discrete mathematics. We are committed to change (as described above), but we expect it to be slow and difficult. Developing a new course and introducing new computer software and discrete viewpoints into traditional calculus are not revolutionary steps, but we hope they will turn out to be feet in the door.

Summary and Conclusions

The goal of the Sloan-supported project at St. Olaf was to develop a curriculum more evenly balanced between discrete and continuous mathematics. We think that our program has begun that process, and that the climate is favorable for the process to continue.

The primary change in the curriculum is the addition of the new discrete mathematics course. The objectives of the course are to introduce students to counting, modular arithmetic, and graph theory and to develop their ability to solve problems, design algorithms, and write proofs. The intended audience is sophomores (or freshmen with advanced placement) who are interested in mathematics or computer science. The course is comparable in difficulty to linear algebra and has a prerequisite of one semester of calculus, in order to insure some mathematical maturity. Student evaluations and increasing enrollment in the course indicate that the course is well received by students. Faculty who have taught the course believe that the content is interesting and worthwhile and that it provides an excellent setting in which to develop problem solving and proof writing skills.

In the calculus and linear algebra course, the main obstacle to including new topics and viewpoints is lack of time. We expect that change will occur in these courses, but recognize that it will occur slowly. We believe that experimentation is worthwhile, and we intend to continue what we have begun.

Course Outline for Discrete Mathematics

Section 0—Sets and functions (1 day)
Review set notation and operations, definition of function, and binary numbers.

PART I. INTRODUCTION TO
PROBLEM SOLVING
AND ALGORITHMS

Section 1—Patterns (1 day)
Demonstrate how examining a few simple cases

of a problem may enable one to detect patterns and then formulate a conjecture.

Section 2—Algorithms (1-2 days)
Describe the pseudocode used for writing algorithms and show how to trace an algorithm.

Section 3—Mathematical induction (2 days)
Introduce the Principle of Mathematical Induction and use it to prove equalities and inequalities. Illustrate the use of induction in algorithm verification.

Section 4—Iteration and recursion (1 day)
Describe the use of iteration and recursion in designing algorithms.

Section 5—Trees and backtracking (1 day)
Introduce trees and backtracking as a method of systematically listing all elements of a set.

PART II. COUNTING

Section 1—Fundamental counting principles (1 day)
State and give examples of addition and multiplication principles of counting. Define probability in the case of equiprobable events.

Section 2—Arrangements (1 day)
Use multiplication rule to solve problems about arrangements of objects chosen from a set with and without repetition. Define permutation and develop formula for number of permutations.

Section 3—Combinations (1 day)
Define combination, prove formula for number of combinations, and solve simple problems involving combinations. Prove the Binomial Theorem.

Section 4—More counting problems (2 days)
Solve harder problems than in sections two and three including problems involving both combinations and permutations. Discuss dividing into cases or subproblems as a problem solving strategy.

Section 5—Analysis of algorithms (1 day)
Define "big oh" notation. Apply counting techniques to analyze time requirements of algorithms.

PART III. INTRODUCTION TO
MATHEMATICAL PROOF

Section 1—Propositional logic (1-2 days)
Present logical symbolism and use it to translate statements into symbolic form. Show how to construct truth tables and use them to test the equivalence of logical formulas and the validity of arguments.

Section 2—Quantifiers (1 day)

Present notation for quantifiers and rules for negating quantified statements. Discuss use of examples in proving and disproving statements.

Section 3—Direct proof (2 days)

Describe the form of a direct proof. Define divides, prime, ceiling and floor functions, and one-to-one and onto functions. Give examples of direct proofs involving these concepts and set properties.

Section 4—Indirect proof (1-2 days)

Describe the form of contradiction and contrapositive proofs. Give examples of indirect proofs involving the concepts introduced in section 3.

Section 5—Induction revisited (1 day)

Show how to use induction to prove statements that are not equalities nor inequalities. State and illustrate strong induction.

Section 6—Pigeonhole Principle (1 day)

State and show how to apply the Pigeonhole Principle.

PART IV. MODULAR ARITHMETIC

Section 1—Congruences (1 day)

Define and prove basic properties of the congruence relation. Use properties of the relation to do modular arithmetic computations, including algorithm for computing powers.

Section 2—Euclidean algorithm and linear congruences (2 days)

Present the Euclidean algorithm and use it to solve congruences and linear diophantine equations. Prove results about existence and number of solutions of congruences.

Section 3—Chinese Remainder Theorem (1 day)

Prove the Chinese Remainder Theorem. Describe algorithms for solving systems of congruences.

Section 4—Applications of modular arithmetic (1-2 days)

Use modular arithmetic to prove divisibility tests and to construct Latin squares. Can also cover public key cryptosystems if time permits.

PART V. GRAPH THEORY

Section 1—Degrees and isomorphism (1 day)

Define graph, degree, subgraph, complement. Prove theorem on sum of degrees. Demonstrate how to show graphs are or are not isomorphic.

Section 2—Paths (1-2 days)

Define path, cycle, connected. Prove theorem on existence of Euler cycle and describe algorithm for finding Euler cycle. Present problems whose solution involves finding an Euler cycle, e.g. DeBruijn sequences.

Section 3—Hamilton paths (1 day)

Demonstrate various methods for showing no Hamilton cycle exists. Describe backtracking algorithm for finding Hamilton path and comment on lack of "good" algorithm.

Section 4—Trees (1-2 days)

Define tree and prove theorem on number of edges in tree. Prove theorem on existence of spanning trees. Use trees to analyze games and solve puzzles.

Section 5—Coloring (1 day)

Define chromatic number and determine chromatic number for small graphs. Present problems whose solution involves coloring, e.g. scheduling.

Section 6—Optimization (2 days)

Describe some of the following algorithms: Kruskal's or Prim's algorithm for minimal spanning tree, Dijkstra's algorithm for shortest path, greedy algorithm for approximating chromatic number, augmented path algorithm for maximal matching.

PART VI. SEQUENCES AND COUNTING

Section 1—Recurrence relations (1 day)

Set up recurrence relations to solve counting problems.

Section 2—Solving recurrence relations (1 day)

Demonstrate how to use iteration to obtain closed form solution of some recurrence relations. Describe method for solving second order linear homogeneous relations.

Section 3—Binomial identities (1-2 days)

Use combinatorial arguments to verify identities involving binomial coefficients. Investigate Pascal's triangle. Use binomial identities to evaluate sums.

Section 4—Calculus of finite differences (1 day)

Introduce the concept of the difference operator and derive some of the difference calculus analogs to differential calculus.

Comments on Course Outline for Discrete Mathematics

The first part is designed to set the tone for the course by encouraging students to explore problems and formulate conjectures. Because of the importance of algorithms in discrete mathematics, systematic approaches to problem solving are emphasized. The sections on induction, iteration and recursion introduce three fundamental problem solving methods which are used throughout the course.

The objective of part two is to develop problem solving skills in the context of the basic concepts of permutation and combination. Elementary combinatorics problems are good vehicles for teaching problem solving for several reasons. First, students often can work with simple cases to get a feel for a problem. Second, the solutions of many combinatorics problems require breaking the problem into a series of simpler subproblems or disjoint cases. Third, since many problems can be solved in more than one way, students can compare the elegance of different solutions. The importance of permutations and combinations in mathematics is illustrated by probability problems and the final section on the analysis of algorithms.

The purpose of part three is to give students practice in writing simple mathematical proofs. Propositional and predicate logic are developed as a means of symbolizing and analyzing mathematical arguments. With this background in logic, the format for direct and indirect arguments can be described. The specific topics treated in the sections on direct and indirect proof - divisibility, sets and functions - were chosen because they are familiar to sophomores and provide many examples of short proofs whose structure is clear. Consequently, instructors can insist on organized, rigorous proofs. The presentation of induction is divided between chapters one and three so that students can gain skill and confidence with induction proofs of equalities before attempting more general induction proofs. The final section on the Pigeonhole Principle relates counting and proof. The principle is used in parts four and five.

The fourth part on modular arithmetic is usually not included as part of a course in discrete mathematics, but we have done so for a number of reasons. The subject provides a setting in which to develop the themes of problem solving, algorithm and proof. Students at all levels of ability can find interesting problems. The algorithms for solving congruences are easily described and useful. The proofs of congruence properties build on the results on divisibility proved in part three. Moreover, experience with modular arithmetic is excellent preparation for the study of quotient groups and finite fields in abstract algebra. The application to cryptography provides a nice case study in the evolution of mathematical ideas and the interplay of pure and applied mathematics.

The fifth part is an introduction to graph theory. Graphs provide an example of a mathematical structure which is not a number system, demonstrate the value of drawing pictures to reflect relationships, and illustrate mathematical modeling. In addition, graph theory is a good setting in which to work with algorithms, particularly optimization algorithms, and gain further practice in proof techniques. Graph theory also provides a context for discussing open questions in mathematics. For example, there are no known necessary and sufficient conditions for the existence of Hamilton cycles, and many graph theory problems are NP-complete.

The final part presents a few topics from combinatorics that are related to sequences. This part uses the counting techniques introduced in part two and requires greater sophistication and creativity. In the section on recurrence relations, recursive thinking is applied to counting problems. The section on solving recurrence relations motivates the study of binomial identities in section three since the solution of a recurrence relation by iteration often reduces to finding a formula for a finite sum. The combinatorial proofs of binomial identities demonstrate how combinatorial arguments may provide an alternative approach to what appear to be algebraic problems. The last section on finite differences relates discrete mathematics and differential calculus.

A Calculus Course With Computer Symbolic Manipulation

The course described below—"Applied Calculus"—was taught in January 1985 and again in January 1986, during St. Olaf's one-month Interim semester. (Students take only one course, meeting about 10 hours per week.) The prerequisite is first semester calculus; no student had more than two semesters. The course has two main features: its emphasis on discrete aspects of calculus (approximation, numerical solution of equations, numerical integration, sequences and recursion, error estimation, etc.), and heavy daily use, by every student, of the computer algebra system SMP. This report describes the course and how symbolic computation figured in it.

Texts: Purcell and Varberg, Calculus, St. Olaf Computer Center Document 920: *Introduction to the Symbolic Mathematics Package SMP*, problem sets produced under Sloan grant, daily handouts, a few excerpts from various sources.

List of Topics

1. Review of "exact" calculus: Overview of first semester material through fundamental theorem.

How SMP can solve canonical calculus problems in "closed form". For comparison, approximation (graphical, numerical) of integrals lacking elementary closed-form solutions—$\exp(x^2), \sin(x^2)$, etc.

2. SMP itself: Syntax of commands. Generic variables. Function definitions. Operations on functions, especially composition, using SMP. Odd and even functions. External files.

3. Polynomial approximations to non-polynomial functions: transcendental vs. explicit algebraic functions, approximate evaluation of transcendental functions. Interpolation by linear, quadratic, piecewise-linear and piecewise-quadratic functions. Lagrange interpolating polynomials of degree n. SMP's polynomial interpolation command Itp. Cubic splines.

4. More polynomials: Taylor and Maclaurin approximations to transcendental functions. Comparing Taylor and interpolation methods of polynomial approximation. Applications of Taylor and interpolation methods: modelling, one-dimensional mechanics—where the phenomena modelled are honestly polynomial functions. Graphs of functions and their polynomial approximations. The SMP function Ps (finds Taylor polynomials).

5. Numerical solution of equations: Bisection method. The idea of an algorithm. Pseudocode. Applications of the bisection method to max-min problems, careful graphing. SMP's Bisol and Bisect functions (the latter exhibits successive approximations). Newton's method.

6. Error estimation: Taylor's theorem with remainder. Error estimates for Newton and bisection methods. Introduction to numerical integration. Graphical techniques for bounding higher derivatives of functions.

7. Numerical integration: Four methods (upper and lower sums, trapezoid, midpoint, Simpson's rules) compared. Error estimate formulas for each. Development of Simpson's Rule from earlier work on quadratic interpolation. Class projects on comparing efficiency and accuracy of various methods. Comparing methods of numerical integration: error estimate formulas, observation that error estimates may overestimate error. Simpson's Rule as weighted average of midpoint and trapezoid rules. Proof of midpoint error formula.

8. Interest. Mathematics of finance. Nominal vs. effective rates. Interest computations for amortization problems. Exponential growth and decay as generalized "interest". Continuous vs. discrete compounding. Applications of numerical tech-

niques: equation-solving, numerical integration in variable-interest-rate setting.

9. First order differential equations. Modelling applications (generalized growth problems). Closed-form techniques (separation of variables, integrating factors). Approximate and numerical methods: polynomial approximation (method of undetermined coefficients), graphical solutions, Euler and Runge-Kutta methods. Error estimates.

10. Sequences: Meaning(s) of convergence—epsilon-N, agreement to fixed number of decimal places. Problem set *Sequences and Their Limits*. Emphasis on recursively defined sequences, e.g., Babylonian square roots.

Conclusion:

SMP worked well, and students enjoyed using it. There was very little trouble with SMP's syntax, perhaps because: the local document seems to be clear; the on-line help system helps; trial-and-error is quicker than asking. In particular:

1. Formal manipulation was reliable and quick, so students could concentrate on concepts. E.g., error estimate formulas involve extrema of higher-order derivatives. For interesting functions, these numbers are hard—and to some degree, pointless—for freshmen to compute with pencil and paper. SMP speeds up these manipulations so students can concentrate on the error estimate idea itself.

2. Students can do more (and more interesting problems) when the data don't have to be expurgated to assure that the equations are quadratic, the integrals have closed-form solutions, etc. With SMP, more of the problems that are solvable in principle are solvable in practice.

3. SMP offers everything the old calculus-by-computer courses did, without the distraction of writing and debugging programs. With SMP, one can teach calculus, not computer programming.

4. Being able to graph, quickly, any function over any interval is surprisingly helpful, even at low resolution. Thus, one can: see that a differentiable function looks straight on a small enough scale, get quick and dirty extremum estimates, roughly locate roots, estimate definite integrals, graph a function and several approximations on one set of axes, etc.

5. SMP forces students to work carefully—commands have to be given precisely. Experiments and trial and error are easy to carry out, but it is also rewarding and easy to work systematically, i.e., to look for algorithms.

6. SMP has many algorithmic procedures built in. More important, it's very easy to implement algorithms in SMP code. For example, SMP evaluates recursively defined functions; Newton's method takes only two lines of "program". (Students did not 'program" in SMP, but with 1–2 extra class days invested, they might have written their own programs for numerical integration, bisection method, Maclaurin polynomial computations, etc.)

7. Students can quickly work through specific examples that illustrate general theorems. Examples: Compute, graph, and tabulate values of several Taylor polynomials. Observe that the errors are as claimed. Find counterexamples to naive differentiation rules. Watch subsequences converge to same limit as sequence.

8. With SMP students experiment, make conjectures, generally work actively with functions—and are rewarded for their efforts. SMP doesn't prove theorems, but it might help clarify the distinction between conjecture and proof, example and principle, estimate and exact value.

Questions and reservations:

1. Does using SMP inhibit students from learning routine computations on their own?

2. Do we have the computer resources to support large-scale use of SMP?

3. If using SMP means push the button and get the answer, then it is worse than useless. For many problems of freshman calculus, this is a risk. Are we willing to spend our own time finding or writing more interesting problems, and class time covering them?

Sample Exercises for Discrete Mathematics

Part I

1. After examining a few special cases, make a conjecture about an expression in terms of n which is equal to each of the following, for $n \geq 1$.
   a. $\frac{1}{1\times 2} + \frac{1}{2\times 3} + \cdots + \frac{1}{n(n+1)}$
   b. $1 + 3 + 5 + \cdots + (2n - 1)$
   c. $[1 - 4/1][1 - 4/9] \cdots [1 - 4/(2n - 1)^2]$

2. The set $\{1, 2, 3\} = \{1, 2\} \cup \{3\}$ and $1 + 2 = 3$. The set $\{1, 2, 3, 4\} = \{1, 4\} \cup \{2, 3\}$ and $1 + 4 = 2 + 3$.

a. For what values of n is it possible to divide $\{1, 2, \ldots, n\}$ into two disjoint subsets such that the sum of each subset is the same?

b. Describe a procedure for dividing $\{1, 2, \ldots, n\}$ into two disjoint subsets satisfying the condition in part a for those integers n for which it is possible.

3. Which positive integers can be expressed in the form $3x + 5y$, where x and y are nonnegative integers?

4. The following algorithm computes the sum of the squares of the first n positive integers, $n \geq 1$.

```
PROCEDURE Sum(n; s)
    *Input: n, a positive integer
    *Output: s, sum of 1² + 2² + ··· + n²
    s ← 0
    FOR i ← 1 to n DO
        s ← s + i²
    ENDFOR
ENDPROC
```

Modify this algorithm to compute the following, for $n \geq 1$.
   a. $1^2 \times 2^2 \times \cdots \times n^2$
   b. $1 + 3 + 5 + \cdots + (2n - 1)$

5. The following algorithm computes the remainder when one positive integer is divided by another positive integer.

```
PROCEDURE Remainder (num,den; r)
    *Input: num, a positive integer
            den, a positive integer
    *Output: r, remainder when num is divided
             by den
    WHILE num ≥ den DO
        num ← (num − den)
    ENDWHILE
    r ← num
ENDPROC
```

Modify this algorithm to compute the quotient as well as the remainder.

6. The following algorithm computes the square of a positive integer.

```
PROCEDURE Square (n; s)
    *Input:  n, a positive integer
    *Output: s, equal to n²
    s ← 0
    d ← 0
    WHILE d < n   DO
        s ← s + n
        d ← d + 1
    ENDWHILE
ENDPROC
```

Trace the algorithm for $n = 4$. Give the values of s and d on each iteration of the loop.

7. Verify your conjectures in Problem #1 by mathematical induction.

8. Use mathematical induction to prove the following, for $n \geq 1$.
   a. $1/\sqrt{1} + 1/\sqrt{2} + \ldots + 1/\sqrt{n} \geq \sqrt{n}$
   b. $[1 \cdot 3 \ldots (2n - 1)]/[2 \cdot 4 \ldots (2n)] < 1/\sqrt{(n + 1)}$

9. In Problem #6, let s_k and d_k be the values of s and d, respectively, after the loop has been executed k times. Verify by induction that $s_k = n \times d_k$ for $k \geq 0$. Show that when the algorithm terminates, $s = n^2$.

10. The following recursive algorithm computes the sum of the first n positive integers.

    PROCEDURE Rsum$(n; s)$
       *Input: n, a positive integer
       *Output: s, equal $1 + 2 + \ldots + n$
       IF $n = 1$
         THEN
           $s \leftarrow 1$
         ELSE
           EXECUTE Rsum$(n - 1; s)$
           $s \leftarrow s + n$
       ENDIF
    ENDPROC

    Modify this algorithm to compute the following.
    a. $n!$, for $n \geq 0$
    b. $1 + 3 + 5 + \ldots + (2n - 1)$, for $n \geq 1$

11. The Fibonacci sequence is the sequence 1, 1, 2, 3, 5, ... Starting with the third term, each term in the sequence is the sum of the two preceding terms. Without using an array, write an algorithm to compute the nth Fibonacci number, where n is a positive integer,
    a. using iteration,
    b. using recursion.

12. Consider the following algorithm.

    PROCEDURE Guess $(d; b)$
       *Input: d, a nonnegative integer
       *Output: b, a sequence of 0's and 1's
       IF $d < 2$
         THEN
           $b \leftarrow d$
         ELSE
           quo $\leftarrow$ quotient when d is divided by 2
           rem $\leftarrow$ remainder when d is divided by 2
           EXECUTE Guess (quo;b)
           $b \leftarrow b$ followed by rem
       ENDIF
    ENDPROC

    a. What is the output if the initial value for d is 26?
    b. What is the purpose of the algorithm?

13. Use a tree to enumerate all arrangements (x_1, x_2, x_3, x_4) of the digits $1, 2, 3, 4$ for which $x_i \neq i$.

14. A basketball squad is made up of three freshmen, four sophomores, and three juniors. Use backtracking to make a list of the possible class distributions for a team of five chosen from this squad. For example, a team might consist of one freshman, two sophomores, and two juniors.

PART II

15. Consider the notes of the scale: C, D, E, F, G, A, B, C'', where C and C'' are low C and high C, respectively. How many eight note melodies, say eight quarter notes, can be composed if
    a. the only allowable notes are C and G?
    b. the first and last notes are either C or G?
    c. repetitions are not allowed?
    d. repetitions are not allowed and G follows immediately after C?
    e. repetitions are not allowed and the first four notes are chosen among C, E, G, B, and the last four notes are among D, F, A, C'?

16. In the previous problem, how many chords can be produced
    a. that consist of four notes?
    b. that consist of three notes and the highest note is A?
    c. that consist of four notes and contain one, but not both, of the C's?

17. Find the constant term (the term free of x) in the expansion of $(3x^2 - 2/x)^{36}$.

18. In Problem #15, how many eight note melodies (eight quarter notes) can be composed if repetition is not allowed and
    a. D must follow A (not necessarily immediately after)?
    b. A and B must be among the first four notes?

19. In Problem #15, suppose repetition is allowed and
    a. the notes are chosen among C, G, and C', with G occurring exactly three times.
    b. A occurs exactly twice among the first five notes and exactly once in the last three.
    c. three of the notes are from C, D, and E, and five of the notes are from F, G, A, B.

20. Show that $\log(n!) = O(n \log n)$.

PART III

21. Let $A = \{3n+1 | n \text{ is an integer}\}$, $B = \{4n+1 | n \in A\}$, and $C = \{12n-7 | n \text{ is an integer}\}$. Prove that $B = C$.

22. Let p, m, and n denote positive integers. Use truth tables to determine if the following argument is valid.

    If p divides the product mn, then either p divides n or p divides m. It is not the case that p divides n. Therefore, if p does not divide m, then p does not divide mn.

23. Write the following statement in symbolic form.

    For all pairs of real numbers x and y with $x < y$ there is a real number z such that $x < z < y$.

    Negate the symbolic expression and then translate the negation into verbal form.

24. Let a, b, c, and d denote integers. One of the following is true and one is false. Prove the one that is true and give a counterexample to the one that is false.

    Suppose that a divides $b+c$ and a divides $c+d$. Then (i) a divides $b - d$; (ii) a divides $b + d$.

25. Decide on the truth of each of the following statements. If true, give a proof; if false, give a counterexample.

    a. For all positive real numbers x, $\lceil x / \lceil x \rceil \rceil = 1$.

    b. For all positive real numbers x, $\lfloor \lceil x \rceil / x \rfloor = 1$.

26. Prove that $x^4 + 1$ cannot be the product of $x^2 + ax + b$ and $x^2 + cx + d$, where a, b, c, d, are integers.

27. Prove that a square can be dissected into k squares (not necessarily congruent squares) for any integer $k \geq 6$.

28. How large a set of distinct integers between 1 and 25 (inclusive) is needed to assure that two numbers in the set will have a common divisor (larger than 1)?

29. The population of a city is 15,363. Each person is asked to give preferences regarding ten possible options. For each option, numbered consecutively from 1 to 10, the person is asked to respond with a "1" (approve), "2" (neutral), or "3" (disapprove). Also, they are instructed to use the approve vote exactly three times. Thus, each completed ballot will consist of a sequence of length ten from the "alphabet" $\{1, 2, 3\}$ in which "1" occurs exactly three times.

    Prove that there are at least four people, A, B, C, D such that A and B cast exactly the same ballot, and C and D cast exactly the same ballot.

PART IV

30. Use congruences to prove the following.

    a. $4^{n+2} + 5^{2n+1}$ is divisible by 21 for every integer $n \geq 1$.

    b. The sum of the squares of three consecutive integers cannot be a square.

    c. $n^5 + 4n$ is divisible by 5 for every integer n.

31. Let p be a prime and $M = \{1, 2, 3, \ldots, p-1\}$, and let $Q = \{a^2 (\bmod p) | a \in M\}$.

    a. By examining special cases, make a conjecture regarding the number of distinct elements in Q.

    b. Prove your conjecture.

32. Prove that for any set of 5 integers, there is a subset of them whose sum is divisible by 5.

33. Find integers x and y such that $41x - 73y = 1$.

34. Solve the following system of congruence equations.

$$3x \equiv 1 (\bmod 5)$$
$$2x \equiv 3 (\bmod 9)$$

35. Let $N = a_n a_{n-1} \ldots a_2 a_1 a_0$ be the decimal representation of the integer N. Prove that N is divisible by 11 if and only if the alternating sum of its digits, that is $a_0 - a_1 + a_2 + \cdots + (-1)^n a_n$, is divisible by 11.

PART V

36. What is the largest number of vertices in a graph with 17 edges if all the vertices have degree at least 4?

37. Consider the two following graphs, G_1 and G_2.

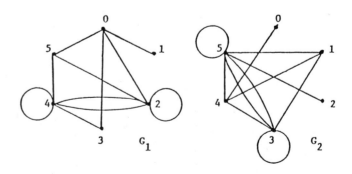

Show that it is possible to "relocate" one of the arcs of G_2 so that the resulting graph will be isomorphic to G_1. Describe the isomorphism by indicating how the vertices of the graphs correspond.

38. For each of the graphs G_1 and G_2 (of Problem #37), describe an Euler path or prove that none exists.

39. A mathematics exam consisted of 28 problems. Each student solved exactly seven problems. For each pair of problems, there were exactly two students who solved both of them. How many students took the exam? (Hint: Let X denote the set of all students and Y the set of all pairs of problems. Construct a graph whose vertices are $X \cup Y$. Draw an edge between $x \in X$ and $y \in Y$ if student x solved both problems in the pair labeled y. Count the edges in this graph in two different ways.)

40. Prove that if a graph has p vertices and every vertex has degree greater than or equal to $(p-1)/2$, then the graph is connected.

41. Find a circular arrangement of the 28 distinct dominoes so that adjacent ends of the dominoes match. (Hint: Show how to translate this problem into an Euler cycle problem.)

42. Criticize the proof given below that purports to show that the given graph does not have a Hamilton cycle.

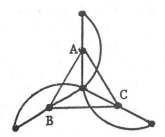

"Proof(?)" The graph has 7 vertices so a Hamilton cycle will have 7 edges. Such a cycle can use at most 2 of the edges at A, at most 2 from B, and at most 2 from C. Therefore, at least 6 edges of the graph won't be part of a Hamilton cycle. The graph has 12 edges, and if 6 of these are not used, this leaves only 6 edges for the cycle. But from above, we need 7 edges. Therefore, there is no such cycle.

43. Does the graph in Problem #42 contain a Hamilton cycle? Prove or disprove.

44. A box weighs x ounces, where x is an integer value between 1 and 16 inclusive. Given a pan balance and weights of 1, 2, 4, and 8, show that the weight of the box can be determined in four or fewer weighings.

45. Two players, A and B, alternately remove stones from a pile which originally contains five stones. Each player removes one or two stones on each move. Player A starts; the player to take the last stone loses. Can A force a win?

46. Seven groups of carolers plan to visit the dorms as follows:

Group1	Group2	Group3	Group4	Group5	Group6	Group7
Kittelsby	Ellingson	Kildahl	Thorson	Thorson	Rand	Ytterboe
Larson	Mellby	Mohn	Kittelsby	Ytterboe	Mohn	Thorson
Mellby	Hilleboe	Rand		Hilleboe		Mohn
	Hoyme			Larson		

Can these visits be scheduled on Friday, Saturday, and Sunday evenings, if no dorm is to be visited more than once in an evening? Model this problem with a graph and show how the problem reduces to finding its chromatic number. If three evenings suffice, what is a satisfactory schedule?

47. Suppose the numbered edges in the following map represent traveling times (in hours between cities). Apply Dijkstra's Algorithm to find the quickest route from S to T.

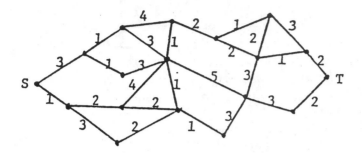

48. In the preceding map, suppose the labels on the edges represent the cost (in millions of dollars) of paving the roads between cities. What is the minimal amount required so that it is possible to travel between any two cities on paved roads (perhaps by way of other cities)?

PART VI

49. Suppose you invest \$50 each month at 10% compounded monthly. How much will you have accumulated at the end of 30 years? (Find a recurrence relation for P_n, the amount accumulated after n months.)

50. Find a recurrence relation for the number of n-digit binary sequences which do not contain the pattern 001.

51. Use a combinatorial argument to prove that

$$\binom{n}{0}\binom{m}{k} + \binom{n}{1}\binom{m}{k-1} + \binom{n}{2}\binom{m}{k-2}$$
$$+ \cdots + \binom{n}{k}\binom{m}{0} = \binom{n+m}{k}$$

52. Prove that the product of the six binomial coefficients that "surround" $\binom{n}{k}$ in Pascal's Triangle, $n > 1$, $0 < k < n$, is a perfect square.

53. Use binomial identities to sum the following.
   a. $\binom{n}{1} + 2\binom{n}{2} + 3\binom{n}{3} + \cdots + n\binom{n}{n}$
   b. $1 \cdot 2 \cdot 3 + 2 \cdot 3 \cdot 4 + 3 \cdot 4 \cdot 5 + \cdots + n(n+1)(n+2)$

54. Use finite differences to find a closed formula expression for the sum of the first n fifth powers,

$$1^5 + 2^5 + 3^5 + \cdots + n^5.$$

Afterthoughts: Discrete Mathematics in the First Two Years

by Martha J. Siegel

The first report of the MAA Committee on Discrete Mathematics appeared in 1985. Already, the textbook list in it is obsolete. There have been at least twenty new books published since then which claim to be written for the freshman-sophomore discrete mathematics market. In addition, many of the books on the committee's list are already out in the next edition. The list that follows this note is an updated bibliography of recently published discrete mathematics books. It will surely be out-of-date by the time this book appears. Information from publishers indicates that the market is still strong for these books although it has settled down somewhat. A great many schools now offer at least a one semester course in discrete mathematics.

Where are we now? The committee was an ad hoc committee, its founding and mission accomplished with the publication of its report. However, the newly formed MAA Committee on the First Two Years of College Mathematics has assumed some of the follow-up tasks. It is hoped that a formal survey of schools will be conducted soon to ascertain the extent to which discrete mathematics is being taught in the first two years of the college curriculum. Questionnaires and telephone interviews were conducted in 1983 by the Committee on Discrete Mathematics.

A BIBLIOGRAPHY OF DISCRETE MATHEMATICS BOOKS INTENDED FOR LOWER DIVISION COURSES

Albertson, Michael O. and Hutchinson, Joan P. DISCRETE MATHEMATICS WITH ALGORITHMS. John Wiley and Sons, 1988. ISBN 0-471-84902-2.

Althoen, Steven C. and Bumbrot, Robert J. INTRODUCTION TO DISCRETE MATHEMATICS. PWS-Kent Publishing Company, 1988. ISBN 0-534-91504-3.

Barnier, William and Chan, Jean B. DISCRETE MATHEMATICS WITH APPLICATIONS. West Publishing Company, 1989. ISBN 0-314-45966-9.

Biggs, Norman L. DISCRETE MATHEMATICS. Oxford University Press, 1985. ISBN 0-19-853252-0.

Bogart, Kenneth. DISCRETE MATHEMATICS. D. C. Heath, 1988. ISBN 0-669-08665-7.

Bradley, James. DISCRETE MATHEMATICS. Addison-Wesley Publishing Company. 1988. ISBN 0-201-10628-0.

Dierker, Paul and Voxman, William. DISCRETE MATHEMATICS. Harcourt Brace Jovanovich, 1986. ISBN 0-15-517691-9.

Doerr, Alan and Levasseur, Kenneth. APPLIED DISCRETE STRUCTURES FOR COMPUTER SCIENCE. Second edition. Science Research Associates, 1988. ISBN 0-574-21755X.

Dossey, John A., Otto, Albert D., Spence, Lawrence E. and Vanden Eynden, Charles. DISCRETE MATHEMATICS. Scott, Foresman, 1987. ISBN 10-673-18191-X

Finkbeiner, Daniel T. and Lindstrom, Wendell D. A PRIMER OF DISCRETE MATHEMATICS. W. II. Freeman and Company. 1987. ISBN 0-7167-1815-4.

Gerstein, Larry. DISCRETE MATHEMATICS AND ALGEBRAIC STRUCTURES. W. H. Freeman and Company, 1987. ISBN 0-7167-1804.

Gersting, Judith L. MATHEMATICAL STRUCTURES FOR COMPUTER SCIENCE. Second Edition. W. H. Freeman and Company, 1987. ISBN 10-7167-1802-2

Grimaldi, Ralph R. DISCRETE AND COMBINATORIAL MATHEMATICS: AN APPLIED INTRODUCTION. Addison-Wesley Publishing Company, 1985. ISBN 0-201-12590-0.

Hillman, Abraham P., Alexanderson, Gerald L., and Grassl, Richard M. DISCRETE AND COMBINATORIAL MATHEMATICS. Dellen Publishing Company, 1987. ISBN 0-02-354580-1.

Johnsonbaugh, Richard. DISCRETE MATHEMATICS, Second Edition. Macmillan Publishing Company, 1987. ISBN 0-02-359690-2.

Johnsonbaugh, Richard. ESSENTIAL DISCRETE MATHEMATICS. Macmillan Publishing Company, 1987. ISBN 0-02-360630-4

Kalmanson, Kenneth. AN INTRODUCTION TO DISCRETE MATHEMATICS AND ITS APPLICATIONS. Addison-Wesley Publishing Company, 1986. ISBN 0-201-14947-8.

Kolatis, Maria Shopay. MATHEMATICS FOR DATA PROCESSING AND COMPUTING. Addison-Wesley Publishing Company, 1985. ISBN 0-201-14955-9.

Kolman, Bernard and Busby, Robert C. INTRODUCTORY DISCRETE STRUCTURES WITH APPLICATIONS. Prentice-Hall, 1984. ISBN 0-13-215418-8.

Kolman, Bernard and Busby, Robert C. INTRODUCTORY DISCRETE STRUCTURES WITH APPLICATIONS. Prentice-Hall, 1987. ISBN 0-13-500794-1.

Marcus, Marvin. DISCRETE MATHEMATICS: A COMPUTATIONAL APPROACH USING BASIC. Computer Science Press, 1983. ISBN 0-914894-38-2.

McEliece, Robert I., Ash, Robert B. and Ash, Carol. INTRODUCTION TO DISCRETE MATHEMATICS. Random House, 1989. ISBN 0-317-58264-X.

Molluzzo, John C. and Buckley, Fred. A FIRST COURSE IN DISCRETE MATHEMATICS. Wadsworth Publishing Company, 1986. ISBN 0-534-05310-6.

Mott, Joe L., Kandel, Abraham and Baker, Theodore P. DISCRETE MATHEMATICS FOR COMPUTER SCIENTISTS. Second Edition. Prentice-Hall. 1986.

Nicodemi, Olympia. DISCRETE MATHEMATICS: A FIRST COURSE FOR STUDENTS OF MATHEMATICS AND COMPUTER SCIENCE. West Publishing Company, 1987.

Norris, Fletcher R. DISCRETE STRUCTURES: AN INTRODUCTION TO MATHEMATICS FOR COMPUTER SCIENTISTS. Prentice-Hall, 1985. ISBN 0-13-215260-6.

Pfleeger, S. L. and Straight, D. W. INTRODUCTION TO DISCRETE STRUCTURES. John Wiley and Sons, 1985. ISBN 0-471-80075-9.

Polimeni, Albert D. and Straight, Joseph H. FOUNDATIONS OF DISCRETE MATHEMATICS. Brooks/Cole Publishing Company, 1985. ISBN 0-535-03612-0.

Prather, Ronald. DISCRETE MATHEMATICAL STRUCTURES FOR COMPUTER SCIENCE. Harghton Mifflin, 1976. ISBN 0-395-20623-7.

Prather, Ronald. ELEMENTS OF DISCRETE MATHEMATICS. Houghton Mifflin, 1986. ISBN 0-395-35165-0.

Roman, Steven. AN INTRODUCTION TO DISCRETE MATHEMATICS. W. B. Saunders, 1986. ISBN 0-03-064019-9.

Ross, Kenneth A. and Wright, Charles R. B., DISCRETE MATHEMATICS. Second edition. Prentice-Hall, 1988. ISBN 0-13-215421-7.

Sedlock, James, T. MATHEMATICS FOR COMPUTER STUDIES. Wadsworth Publishing Company, 1985. ISBN 0-534-04326-7.

Shiflet, Angela B. DISCRETE MATHEMATICS FOR COMPUTER SCIENCE. West Publishing Company, 1987. ISBN 0-314-28513-X.

Skvarcius, Romualdas and Robinson, William B. DISCRETE MATHEMATICS WITH COMPUTER SCIENCE APPLICATIONS. Benjamin Cummings Publishing Company, 1986. ISBN 0-8053-7044-7.

Wiitala, Stephen A. DISCRETE MATHEMATICS: A UNIFIED APPROACH. McGraw-Hill, 1987. ISBN 0-07-070169-5.

Williamson, Gill S. COMBINATORICS FOR COMPUTER SCIENCE. Computer Science Press, 1986. ISBN 0-88175-020-4.

FINAL REPORT OF THE MAA COMMITTEE ON DISCRETE MATHEMATICS IN THE FIRST TWO YEARS

SUMMARY OF RECOMMENDATIONS

1. Discrete mathematics should be part of the first two years of the standard mathematics curriculum at all colleges and universities.

2. Discrete mathematics should be taught at the intellectual level of calculus.

3. Discrete mathematics courses should be one year courses which may be taken independently of the calculus.

4. The primary themes of discrete mathematics courses should be the notions of proof, recursion, induction, modeling and algorithmic thinking.

5. The topics to be covered are less important than the acquisition of mathematical maturity and of skills in using abstraction and generalization.

6. Discrete mathematics should be distinguished from finite mathematics, which, as it is now most often taught, might be characterized as baby linear algebra and some other topics for students not in the "hard" sciences.

7. Discrete mathematics should be taught by mathematicians.

8. All students in the sciences and engineering should be required to take some discrete mathematics as undergraduates. Mathematics majors should be required to take at least one course in discrete mathematics.

9. Serious attention should be paid to the teaching of the calculus. Integration of discrete methods with the calculus and the use of symbolic manipulators should be considered.

10 . Secondary schools should introduce many ideas of discrete mathematics into the curriculum to help students improve their problem-solving skills and prepare them for college mathematics.

INTRODUCTION AND HISTORY

The Committee on Discrete Mathematics in the First Two Years was established in the spring of 1983 for the purpose of continuing the work begun at the Williams College Conference held in the summer of 1982. That conference brought to a forum the issue of revising the college curriculum to reflect the needs of modern programs and the students in them. Anthony Ralston and Gail Young brought together 29 scientists (24 of whom were mathematicians) from both industry and academe to discuss the possible restructuring of the first two years of college mathematics. Although the growing importance of computer science majors as an audience for undergraduate mathematics was an important motivation for the Williams Conference, the conference concerned itself quite broadly with the need to revise the first two years of the mathematics curriculum for everyone - mathematics majors, physical science and engineering majors, social and management science majors as well as computer science majors. The papers presented and discussed at the conference, and collected in *The Future of College Mathematics* [11], reflect this breadth of view.

The word used to describe what was needed was "discrete" mathematics. Most of us knew what that meant approximately and respected the content as good mathematics. To illustrate the discrete mathematics topics that might be considered for an elementary courses, two workshop groups at the Williams Conference produced (in a very short time) a fairly remarkable set of two-course sequences:

1. a two year sequence of independent courses, one in discrete mathematics and one in a streamlined calculus, and

2. a two year integrated course in discrete and continuous mathematics (calculus) in a modular form for service to many disciplines.

These course outlines were admittedly tentative and needed refinement and testing. At the same time, the MAA Committee on Service Courses had been examining the traditional service course offerings of the first two years. The syllabi of these courses, in which many freshman and sophomores are required to enroll, are studied for their relevancy periodically. Finite mathematics, linear algebra, statistics, and calculus are considered to be essential to many majors, but with the importance of the computer, the Committee on Service Courses concluded that even the mathematics majors need mathematics of a new variety, not only so they can take computer science courses, but also so they can work on contemporary problems in mathematics.

At that time, there were few or no textbooks or examples of such courses for the community to share. At the suggestion of the Committee on Service Courses, the MAA agreed to help to develop the Williams courses further through the Committee on the Undergraduate Program in Mathematics (CUPM) and the Committee on the Teaching of Mathematics (CTUM), standing committees of the Association. That led to the establishment of the Committee responsible for this report. Funds for the effort were secured from the Sloan Foundation. The members of the Committee were chosen especially to reflect the communities who would eventually be most affected by any changes in the traditional mathematics curriculum. They are:

Martha J. Siegel, Department of Mathematics, Towson State University, member of CTUM, Chairperson;

Alfs Berztiss, Department of Computer Science, University of Pittsburgh, representative of the ACM Education Board;

Donald Bushaw, Department of Pure and Applied Mathematics, Washington State University, member of CTUM and of CUPM, chairperson of MAA Committee on Service Courses;

Jerome Goldstein, Department of Mathematics, Tulane University, chairperson of CUPM;

Gerald Isaacs, Department of Computer Science, Carroll College, representative of the ACM Education Board;

Anthony Ralston, Department of Computer Science, State University of New York at Buffalo, MAA Board of Governors, an organizer of the Williams Conference;

John Schmeelk, Department of Mathematics, Virginia Commonwealth University, member of The American Society for Engineering Education (ASEE) Mathematics Education Committee;

Stephen Maurer, Department of Mathematics, Swarthmore College, then on leave at The Alfred P. Sloan Foundation.

In addition to the development of course outlines

and plans for their implementation, the Committee was involved in the observation of a set of experimental projects which also were begun as a result of the Williams Conference and the interest of the Sloan Foundation. After the Conference, the Sloan Foundation solicited about thirty proposals for courses which would approximate the syllabi suggested by the workshop participants. The call for proposals particularly mentioned the need for the development of text material and classroom testing and emphasized the hope that some schools would make the effort to try the integrated curriculum (2. above). Six schools were ultimately chosen:

Colby College, Waterville, ME,
University of Delaware, Newark, DE,
University of Denver, Denver, CO,
Florida State University, Tallahassee, FL,
Montclair State College, Montclair, NJ,
St. Olaf College, Northfield, MN.

The Committee, together with some of the committee which chose the proposals to be funded (Don Bushaw, Steve Maurer, Tony Ralston, Alan Tucker, and Gail Young), monitored their progress for the two year period of funding, ending in August, 1985. [Accounts of the results of these projects are the main content of this book.]

GENERAL DISCUSSION OF DISCRETE MATHEMATICS COURSES

In its final report, the Committee has decided to present two course outlines for elementary mainstream discrete mathematics courses. Our unanimous preference is for a one year course, at the level of the calculus but independent of it. It is designed to serve as a service course for computer science majors and others and as a possible requirement for mathematics majors. The Committee on the Undergraduate Program in Mathematics (CUPM) has endorsed the recommendation that every mathematics major take a course in discrete mathematics and has agreed that the year course the Committee recommends is a suitable one for mathematics majors. It is expected that the course will be taken by freshmen or sophomores majoring in computer science so that they can apply the material in the first and second year courses in their major. The ACM recommendations [7,8] for the first year computer science course presume, if not specific topics, then certainly the level of maturity in mathematical thought which students taking the discrete mathematics course might be expected to have attained. Hence, the Committee recommends that the course be taken simultaneously with the first computer science course. The Committee understands that at some schools the first computer science course may be preceded by a course strictly concerned with programming. At the very least, the Committee expects that the discrete mathematics course will be a prerequisite to upper level computing courses. For this reason, the Committee has tried to isolate those mathematical concepts that are used in computer science courses. The usual sequence of these courses might determine what should be taught in the corresponding mathematics courses.

In addition to the Committee's concern for computer science majors, there is a high expectation that mathematics majors and those in most physical science and engineering fields will benefit from the topics and the problem solving strategies introduced in this discrete mathematics course. Subjects like combinatorics, logic, algebraic structures, graphs and network flows should be very useful to these students. In addition, methods of proof, mathematical induction, techniques for reducing complex problems to simpler (previously solved) problems and the development of algorithms are tools to enhance the mathematical maturity of all. Furthermore, students in these scientific and mathematically oriented fields will want to take computer science courses and will need some of the same mathematical preparation that the computer science major needs.

Thus, the Committee has agreed to recommend that the course be part of the regular mathematics sequence in the first two years for all students in mathematically related majors. Our contacts with physicists and engineers reinforce the idea that their students will need this material but, of course, there is the concern that calculus will be short-changed.

The Committee will make several suggestions regarding the calculus, but individual institutions will best understand their own needs in this regard. We do not recommend that the third semester of calculus be cut from a standard curriculum. Serious students in mathematical sciences, engineering and physical sciences need to know multivariable calculus. Many in the mathematical community recognize that the content of the calculus should be updated to acknowledge the use of numerical methods and computers, and promising initiatives along this line are being taken. Engineers have been especially anxious for this change. John Schmeelk surveyed 34 schools and compiled suggestions for revising the standard calculus. These are included in the appendix [see original report]. At some of the Sloan-funded schools and others there have been attempts to revise the calculus to incorporate some discrete methods and to use the power of the symbolic manipulation packages. We describe these attempts later in the report.

There is, inherent in our proposal, the possibility that some students may be required to take five semesters of mathematics in the first two years - a year of discrete mathematics and the three semesters of calculus. But, there is no reason why students cannot be allowed to take one of the five in the junior year. We point out that some linear algebra is included in the year of discrete mathematics. Additionally, the use of computers via the new and powerful symbolic manipulation packages may reduce the amount of time needed for the traditional calculus sequence.

A one-semester discrete mathematics course will be described in the appendix to this report as a concession to the political realities in many institutions. It has become obvious to the Committee over the last two years that, at some colleges, there is a limitation on the number of new elementary courses that can be introduced at this time.

The Committee believes strongly that mathematics should be taught by mathematicians. Although there are some freshman-sophomore courses in discrete mathematics in computer science departments, the course presented here should, the Committee believes, be taught by mathematicians. The rigor and pace of this course are designed for the freshman level. Some topics necessary for elementary computer science may have to be taught at an appropriate later time, either in a junior level discrete mathematics course or in the computer science courses.

NEEDS OF COMPUTER SCIENCE

What do the computer science majors need? In teaching the first year Introduction to Computer Science course, Tony Ralston kept track of mathematics topics he would have liked the students to have had before, or at least concurrently with, his course. The results appear below. Many of the ideas are those that students *should have had* in four years of the traditional high school curriculum. In addition, there are some ideas and techniques that are probably beyond the scope of secondary school mathematics. An elaboration of this list appears in the Appendix and in an article by Ralston in ACM *Communications* [10].

MATHEMATICS NEEDED TO SUPPORT
THE FIRST COMPUTER SCIENCE COURSE*

ELEMENTARY MATHEMATICS

Summation notation

Subscripts

Absolute value, truncation, logarithms, trigonometric functions

Prime numbers

Greatest common divisor

Floor and ceiling functions

GENERAL MATHEMATICAL IDEAS

Functions

Sets and operations on sets

ALGEBRA

Matrix algebra

Polish notation

Congruences

SUMMATION AND LIMITS

Elementary summation calculus
Order notation, $O(f(n))$
Harmonic numbers

NUMBERS AND NUMBER SYSTEMS

Positional notation

Nondecimal bases

LOGIC AND BOOLEAN ALGEBRA

Boolean operators and expressions

Basic logic

PROBABILITY

Sample spaces

Laws of probability

COMBINATORICS

Permutations, combinations, counting

Binomial coefficients, binomial theorem

GRAPH THEORY

Basic concepts

Trees

DIFFERENCE EQUATIONS AND

RECURRENCE RELATIONS

Simple difference equations
Generating functions

*See [10]

Many proposals have been coming from the computer science community. Recommendations for a freshman level discrete mathematics course from the Educational Activities Board of IEEE probably are the most demanding. Students enrolled in the course outlined in the appendix are first semester freshmen also enrolled in the calculus according to the IEEE recommendations published in December, 1983, by the IEEE Computer Society [6].

Accreditation guidelines passed recently by ACM and IEEE also require a discrete mathematics course. The recommendations for the mathematics component of a program that would merit accreditation appear below. The criteria appear in their entirety in an article by Michael Mulder and John Dalphin in the April 1984 *Computer* [9].

Certain areas of mathematics and science are fundamental for the study of computer science. These areas must be included in all programs. The curriculum must include one-half year, equivalent to 15 semester hours, of study of mathematics. This material includes discrete mathematics, differential and integral calculus, probability and statistics, and at least one of the following areas: linear algebra, numerical analysis, modern algebra, or differential equations. It is recognized that some of this material may be included in the offerings in computer science...

Presentation of accreditation guidelines which require one and one-half years of study in computer science, one year in the supporting disciplines, one year of general education requirements and one-half year of electives induced quick and angry response. The liberal arts colleges and the small colleges unable to offer this number of courses or unwilling to require so many credits in one discipline, have responded in many ways. The Small College Task Force of the ACM issued its own report, approved by the Education Board of the ACM [1]. We emphasize only the mathematics portion of those guidelines.

> Many areas of the computer and information sciences rely heavily on mathematical concepts and techniques. An understanding of the mathematics underlying various computing topics and a capability to implement that mathematics, at least at a basic level, will enable students to grasp more fully and deeply computer concepts as they occur in courses... It seems entirely reasonable and appropriate, therefore, to recommend a substantial mathematical component in the CSIS curriculum... To this end, a year of discrete mathematical structures is recommended for the freshman year, prior to a year of calculus.

The Sloan Foundation supported representatives of a few liberal arts schools in their attempt to define a high-quality computer science major in such institutions. Again, we put the emphasis on the mathematics component of the proposed program.

From *Model Curriculum for a Liberal Arts Degree* in Computer Science, by Norman E. Gibbs and Allen B. Tucker [4]:

> The discrete mathematics course should play an important role in the computer science curriculum ... We recommend that discrete mathematics be either a prerequisite or corequisite for CS2. This early positioning of discrete mathematics reinforces the fact that computer science is not just programming, and that there is substantial mathematical content throughout the discipline. Moreover, this course should have significant theoretical content and be taught at a level appropriate for freshman mathematics majors. Proofs will be an essential part of the course.

Alfs Berztiss, a member of the Committee, led a number of mathematics and computer science faculty at a conference at the University of Pittsburgh in 1983 at which an attempt was made to formulate a high quality program in computer science which would prepare good students for graduate study in the field. Details are available in a Technical Report (83-5) from the University of Pittsburgh Department of Computer Science [2]. Both that program and the new and extensive bachelor's program in computer science at Carnegie-Mellon University depend on an elementary discrete mathematics course.

In addition to the proposals for programs, the computer science community is in the process of revising elementary computer science courses. Though old courses stressed language instruction, a more modern approach stresses structured programming and a true introduction to computer science. The beginning course CS1 and CS2 are described by the ACM Task Force on CS1 and CS2. We quote from the article by Elliot Koffman et al. [7] about the role of discrete mathematics in the structure of these computer science courses.

> We are in agreement with many other computer scientists that a strong mathematics foundation is an essential component of the computer science curriculum and that discrete mathematics is the appropriate first mathematics course for computer science majors. Although discrete mathematics must be taken prior to CS2, we do not think it is a necessary prerequisite to CS1. ...We would...expect computer science majors and other students interested in continuing their studies in computer science to take discrete mathematics concurrently with the revised CS1.

If high schools and colleges take the recommendation seriously, the student enrolled in CS1 would be enrolled in a discrete mathematics course concurrently. That mathematics course would be required as a prerequisite for CS2. Of all the recommendations, this is likely to have the largest impact on enrollments in discrete mathematics courses. Readers should consult the appendix [see original report] for more information on these matters.

THE SYLLABUS OF A DISCRETE MATHEMATICS COURSE

What are the common needs of mathematics and computer science students in mathematics? The Committee agrees that all the students need to understand the nature of proof, and the essentials of propositional and predicate calculus. In addition, all need to understand recursion and induction and, related to that, the analysis and verification of algorithms and the algorithmic method. The nature of abstraction should be part of this elementary course. While some of the Committee supported the introduction of algebraic structures in this course, particularly for coding theory and finite automata, others felt that those concepts were best left to higher level courses in mathematics. The basic principles of discrete probability theory and elementary statistics might be considered to be as important and more accessible to students at this level. Professionals in all disciplines cite the importance of teaching problem solving skills. Graph theory and combinatorics are excellent vehicles. All these students need some calculus.

The Committee recommends the inclusion of as many of the proposed topics as possible with the understanding that taste and the structure of the curriculum in each institution will dictate the depth and extent to which they are taught. The ability of students in a course at this level must be considered in making these choices. While one of the goals of the course is to increase the mathematical maturity of the student, some of the mathematical community who have communicated to the panel about their experiences teaching this course have indicated that there are prerequisite skills in reading and in maturity of thinking that really are needed, perhaps even more than in the calculus.

The Committee recognizes that it might be some time before there is as much agreement on the content of a discrete mathematics sequence as there is now about the calculus sequence. In the meantime, diversity and variety should be encouraged so that we may learn what works and what does not. In any case, the Committee strongly endorses the notion that it is not what is taught so much as how. If the general themes mentioned in the previous paragraph are woven into the content of the course, the course will serve the students well. Adequate time should be allowed for the students to DO a lot on their own: they should be solving problems, writing proofs, constructing truth tables, manipulating symbols in Boolean algebra, deciding when, if, and how to use induction, recursion, proofs by contradiction, etc. And their efforts should be corrected.

We have been asked about the role of the computer in this course. To a person we have agreed that this is a mathematics course and that while students might be encouraged, if they have the background, to try the algorithms on a computer, the course should emphasize mathematics. The skills that we are trying to teach will service the student better than any programming skills we might teach in their place, and the computer science departments prefer it that way. Surely the ideal would be that students be concurrently enrolled in this course and a computer science course where the complementary nature of the subjects could be made clear by both instructors.

Algorithms are, of course, an integral part of the course. There is still no general agreement on how to express them in informal language. While a form of pseudocode might suit some people, others have found that an informal conversational style suffices. The Committee would not want to make any specific recommendations except that the student be precise and convey

his/her methods. It is certainly NOT necessary to write all algorithms in Pascal. Communication is the key.

The recommendations for a one-year discrete mathematics course are presented in several ways. An outline of the course appears below. In the appendix [see original report], the outline has been expanded to include objectives and sample problems for each topic. The scope and level of the course can be appreciated best from the expanded version.

PRELIMINARY OUTLINE OF A ONE YEAR FRESHMAN-SOPHOMORE COURSE IN DISCRETE MATHEMATICS

Prerequisite: Four years of high school mathematics; may be taken before, during, or after calculus I and II.

1. Sets

   finite sets, set notation, set operations, subsets, power set, sets of ordered pairs, Cartesian products of finite sets, introduction to countably infinite sets

2. The number system

   natural numbers, integers, rationals, reals, Zn, primes and composites, introduction to operations and algebra

3. The nature of proof

   use of examples to demonstrate direct and indirect proof, converse and contrapositive, introduction to induction, algorithms

4. Formal logic

   propositional calculus, rules of logic, quantifiers and their properties, algorithms and logic, simplification of expressions

5. Functions and relations

   properties of order relations, equivalence relations and partitions, functions and properties, into, onto, 1-to-1, inverses, composition, set equivalence, recursion, sequences, induction proofs

6. Combinatorics

   permutations, combinations, binomial and multinomial coefficients, counting sets formed from other sets, pigeon-hole principle, algorithms for generating combinations and permutations, recurrence relations for counting

7. Recurrence relations

   examples, models, algorithms, proofs, the recurrence paradigm, solution of difference equations

8. Graph Theory

definitions, applications, matrix representation of graphs, algorithms for path problems, circuits, connectedness, Hamiltonian and Eulerian graphs, ordering relations - partial and linear ordering, minimal and maximal elements, directed graphs

9. Trees

binary trees, search problems, minimal spanning trees, graph algorithms

10. Algebraic Structures

boolean algebras, semigroups, monoids, groups, examples and applications and proofs

OR

10. Discrete probability and descriptive statistics

events, assignment of probabilities, calculus of probabilities, conditional probability, tree diagrams, Law of Large Numbers, descriptive statistics, simulation

11. Algorithmic Linear Algebra

matrix operations, relation to graphs, invertibility, row operations, solution of systems of linear equations using arrays, algebraic structure under operations, linear programming - simplex and graphing techniques

PREPARATION FOR DISCRETE MATHEMATICS

A consideration of the topics listed in this course outline reveals that, while the course meets our objectives of scope and level, this is a serious mathematics course. The student will have to be prepared for this course by an excellent secondary school background. Those of us who have been teaching freshmen know that many students are coming unprepared for abstract thinking and problem solving. We are aware that many secondary schools are doing a fine job of educating students to handle this work, but many more schools are not. It seems likely that courses ordinarily taught to mathematically deficient first year students to prepare them for the calculus would also prepare them for this course. In many cases, with only modest changes, these courses can be adapted to be both pre-discrete mathematics and precalculus. The Committee expects that the major in computer science will surely reap the full benefits of these traditional preparatory courses. The additional question still remains unanswered - what should be taught in the high schools or on the remedial level in the colleges to prepare students adequately for this course? Our suggestion is tentative: some of us feel that

perhaps a revived emphasis on the use of both formal and informal proof in geometry courses as a means for teaching methods of proof and analytic thinking would be a step in the right direction. Others of us are not so sure. Increased use of algorithmic thinking in problem solving could be easily adapted to many high school courses. Readers are encouraged to read Steve Maurer's article in the September, 1984 *Mathematics Teacher* for some more on this subject.

The Committee on Placement Examinations of the MAA will be attempting to isolate those skills that seem to be needed by students taking discrete mathematics. Although this study might not lead to the development of a placement examination for the course, it will help to explain what might be the appropriate preparation for a successful experience in such a course.

Year after year we face students who claim that they have never seen the binomial theorem, mathematical induction, or logarithms before college. These used to be topics taught at the eleventh or twelfth grade levels. What has happened to them? Students also say that they never had their papers corrected in high school so they never wrote proofs. Some of us have students who cannot tell the hypothesis from the conclusion.

Simple restoration of some of the classical topics and increased emphasis on problem solving might make the proposed course much easier for the student. As one studies the list of topics in the discrete mathematics course, it becomes clear that, in fact, there is little in the way of specific prerequisites for such a course except a solid background in algebra; nothing in the course relies on trigonometry, number theory, or geometry, *per se*. However, the abstraction and the emphasis on some formalism will shock the uninitiated and the mathematically immature.

Recent experimentation at the Sloan-funded schools might tell us something about what we ought to require of students enrolling in this type of course. Results from these schools have not been completely analyzed, but the failure rates seem consistent with those in the calculus courses. Some of the experimental group had taken calculus first and others had not. There seemed to be a filtering process in both cases so that results are not comparable from one discrete mathematics course to another. One Sloan-funded correspondent reported that reading skill might be a factor in success and was following through with a study to see if verbal SAT scores were any indicator of success.

One of the concerns of the Committee throughout its deliberations has been the articulation problem with a course of this kind. We want to be clear that finite mathematics courses in their present form are not the equivalent of this course. We have not totally succeeded in communicating this in presentations at professional meetings. The discrete and finite mathematics courses differ in several ways. First, the discrete mathematics course is not an all-purpose service course. It has been designed primarily for majors in mathematically-related fields. It presumes at least four years of solid secondary school mathematics and hence the level of the course is greater than or equal to the level of calculus. There is inherent in this proposal a heavy emphasis on the use of notation and symbolism to raise the students' ability to cope with abstraction. Secondly, a heavy emphasis on algorithmic thinking is also recommended.

The pace, the rigor, the language and the level are intended to differ from a standard finite mathematics course. We do not claim that this course can be taught to everyone. Perhaps at some schools, the computer science majors are not very high caliber and college programs naturally are geared to the needs of the students. There is nothing inherently wrong in requiring that such students take the mathematics courses required of the business majors: finite mathematics, basic statistics and "soft" calculus. Perhaps the finite mathematics courses can be improved and sections for some students can be enhanced by teaching binary arithmetic and elementary graphs. This is an alternative that many schools will probably choose. It may reflect the reality on a campus where there is really no major in computer science, but a major in data processing or information science which services its students well. We have not attempted to define that kind of discrete mathematics course. We specifically are defining a course on the intellectual level of calculus for science and mathematics majors. Our visits around the country indicate that many schools need a course at the level of the present finite mathematics offerings. Such courses are a valuable service to some students, but should not be considered equivalent to the course we have described.

DISCRETE MATHEMATICS IN TWO YEAR COLLEGES AND HIGH SCHOOLS

The mathematics faculty at two-year colleges have been working through their own organizations and committees toward curricular reform. The Committee on Discrete Mathematics has attempted to consider their proposals in its own. Jerry Goldstein, Chairperson of CUPM and an ex officio member of the Committee on the Curriculum at the Two Year Colleges, has been working to maintain articulation between the two groups. The Two Year College Committee began its deliberations after our Committee, so this report re-

flects only preliminary conclusions from that source. A "Williams"-like conference for the two year colleges took place in the summer of 1984 and proceedings are available from Springer-Verlag in *New Directions in Two-Year College Mathematics*, edited by Donald Albers et al. The situation at this time in the two year colleges is one of exploration, learning, and waiting.

Just as the calculus sequence at two-year colleges is taught from the same texts and in the same manner as at the baccalaureate institutions, discrete mathematics courses at two year schools are expected to conform to requirements of four year schools to which students hope to transfer. Faculty at Florida State University, in connection with one of the Sloan projects, introduced the discrete mathematics course at a nearby two year college. The course was taught from the same text and in the same manner at both institutions. The students did well and project directors claim the results were "unremarkable."

Recent conferences of the American Mathematical Association of Two Year Colleges (AMATYC) and associations of two year college mathematics faculty in many state organizations have been devoted to the special problems of the two year schools with regard to discrete mathematics. The primary concern of most schools is that they must wait for the four year schools to indicate what type of course will be transferable. The Committee urges those teaching at four year institutions to make a special effort to communicate their own requirements to the two year colleges that feed them.

What about discrete mathematics in the high schools? Perhaps it will be an exciting change to see the secondary schools place less emphasis on calculus and more on some of the topics in the discrete mathematics. We understand that there is considerable pressure from parents to have Junior (or Sis) take calculus in high school. We are confident that that will change as the first year of mathematics in the colleges becomes more flexible to include either calculus or discrete mathematics at the same level. If the high schools continue the trend to teaching more computer science for advanced placement, then they will have to offer the discrete mathematics to their students. The present Advanced Placement Examination in computer science is essentially for placement in CS2. To place above CS2, there will probably be a level II examination which conforms to the course outline for CS1 and CS2 as noted in the Koffman report. [ETS now offers CSA and CSAB.] An Advanced Placement Examination in discrete mathematics is some time in the future, as there is no universal agreement as to exactly what ought to be included.

In January 1986 a Sloan-funded conference on calculus was held at Tulane University in New Orleans [3]. More than twenty participants presented papers and participated in workshops on the state of calculus and its future. The Committee concurs with that groups' consensus that the goals of teaching (mathematics) are to develop increased conceptual and procedural skills, to develop the ability of students to read, write and explain mathematics, and to help students deal with abstract ideas. These are the global concerns for all mathematics teaching. Secondary schools should be working toward such goals too. The Committee encourages faculty to get students to work together to solve problems. From experience, some of us have found that students cannot read a problem - either they leave out essential words or do not know how to read the notation when asked to read aloud. The word "it" should be banned from their vocabulary for a while. Students who use the word frequently do so because they do not know what "it" really is. Correcting students' homework has always been one of the best ways of understanding their misconceptions. In discrete mathematics courses this is even more so - concept and procedure vary from problem to problem. Students have to think and be creative. That's tough. They need the reinforcement of the teacher's comments and the chance to try again. Working with other students should be encouraged because this forces students to speak. This oral communication helps them to learn the terminology and helps them to present clear explanations.

THE IMPACT ON THE CALCULUS

The concerns of some people that the introduction of discrete mathematics will cause a major change in the calculus will probably prove to be unfounded. However, the Committee believes that there are several important questions to be addressed. We should be asking ourselves if we are doing the best job of teaching calculus. Some of our colleagues outside of mathematics who teach our calculus students have commented to the Committee members that there are many aspects of the calculus which seem to be ignored in the present courses. There is widespread dissatisfaction with the problem solving skills of calculus students. Problems that look even a little different from the ones that they have solved in the standard course are often impossible for students. In addition, we are being held responsible for our students lack of knowledge of numerical techniques. The discrete aspect of the calculus was continually stressed by our respondents. In fact, many commented that we were promoting the idea of a dichotomy in mathematics where there is none by not proposing an integrated program of discrete and continuous mathematics for the first two years. The Committee admits that at this time it is presenting a feasible solution as opposed to the ideal solution.

Should the teaching of calculus reflect the tremendously powerful symbolic manipulators now on the market? While the most powerful require mainframes, some are available on minicomputers and muMath runs on a personal computer. Can the time previously spent in tedious practice of differentiation or integration be better used to teach the power of the calculus through problem solving and modeling? Two of the Sloan-funded schools - Colby College and St. Olaf College - did experiment with the use of MACSYMA, MAPLE and SMP in the teaching of calculus. At Colby, in a course offered to those who had high school calculus, the computer packages were used to augment the one year single-variable and multivariate calculus course. At St. Olaf, SMP was used in an elective course during a January interim between the first and second semesters of the standard calculus course. Kathleen Heid writes in *The Computing Teacher* [5] about her experience at the University of Maryland where she taught a section of the "soft" calculus using muMath. The results of all these experiments are quite favorable and indicate an important new consideration in our teaching of the subject.

What about the use of the methods introduced in discrete mathematics in the other courses in the curriculum, including calculus and analysis? What of difference equations? The Committee requested that physical scientists and engineers respond to the idea of changing the calculus. We mentioned the possibility that calculus might contain ideas from discrete mathematics in the solving of traditional calculus-type problems. Several engineers and physicists have responded to our query with some interesting endorsements for change. Those who responded felt that the present mathematical training we offer their professions is inconsistent with what many of them were doing in their jobs - for they were using difference equations and other discrete methods in their everyday applications.

We also should be asking what calculus the computer science major needs. Does the computer science student need the calculus to do statistics and probability? If so, how much rigor is needed? What background is needed in numerical methods? Should mathematics departments be teaching numerical methods? Are the requirements different from numerical analysis? Should we emphasize rigor, technique or problem solving skills? Do the traditional courses suffice to encourage integration of discrete and continuous mathematics?

CONCLUSION

This report is both incomplete and already out-of-date. Questions will continue to arise; answers are not easily found. Textbooks are now being published that are marketed as suitable for elementary discrete mathematics courses. Our annotated bibliography is undoubtedly incomplete. We know of several forthcoming texts that are in manuscript form but which are unlisted because they could not be properly reviewed.

There has been a great deal of interest, much of it enthusiastic, in the revitalization of the elementary college-level mathematics curriculum. The Committee members have had the opportunity to visit schools, speak at sectional and national meetings, and to speak personally with hundreds of our colleagues. We are wrestling with problems of ever-changing demands from other disciplines - some, as computer science, so young there is no standard curriculum. We need to adjust our ideals to the realities of our own academic situation. The Committee attempted to propose a course with enough flexibility to allow institutions with different needs to follow the general course outline, putting emphases where they wanted.

The two year colleges and the high schools are dealing with demands of the four year institutions, parents and the College Entrance Examination Board. They feel many pressures to keep calculus as the pivotal course. On the other hand, the proposal to integrate discrete mathematics into the high school and even elementary school curricula got considerable support at the 1985 National Council of Teachers of Mathematics (NCTM) meetings in San Antonio. [The 1989 NCTM *Standards* include this.]

The recent publication of many discrete mathematics textbooks suitable for the freshman-sophomore year has been exciting. We have the opportunity to see what is successful. The Committee agrees that the next step in the development of the curriculum should be the integration of the discrete and the continuous ideas of mathematics into all courses. That would be ideal and we encourage experimentation to that end.

REFERENCES

1. Beidler, John; Austing, Richard H., Cassel, Lillian N. Computing programs in small colleges. *Communications of the ACM*; 1985; 28(6): 605-611. Note: Summary report of The ACM Small College Task Force; outlines resources, courses, and problems for small colleges developing degree programs in computing. See especially "The Mathematics Component," p. 610.

2. Berztiss, Alfs T. Towards a rigorous curriculum for computer science. Technical Report 83-5, University of Pittsburgh Department of Computer Science; 1983.

3. Douglas, Ronald G. (Ed). *Toward a Lean and Lively Calculus* (Conference/Workshop to Develop Alternative Teaching Methods for Calculus at the College Level). Washington: Mathematics Association of America; 1986:; ISBN 0-88385-056-7.

4. Gibbs, Norman E.; Tucker, Allen B. Model curriculum for a liberal arts degree in computer science. *Communications of the ACM*; 1986; 29(3). Note: A curriculum developed by computer scientists supported by a grant from the Sloan Foundation. The purpose was to define a rigorous undergraduate major in computer science for liberal arts colleges.

5. Heid, M. Kathleen. Calculus with muMath: Implications for curriculum reform. *The Computing Teacher*; 1983; 11(4): 46-49.

6. IEEE Educational Activities Board, Model Program Committee. The 1983 IEEE Computer Society model program in computer science and engineering. New York: The Institute of Electrical and Electronics Engineers; 1983. Note: A section "Discrete mathematics" appears on pp. 8-12. This contains a rather demanding modular outline of topics to be covered, and urges integration of the mathematical theory with computer science and engineering applications.

7. Koffman, E.; Miller, P.; Wardle, C. Recommended curriculum for CS1, 1984: A report of the ACM Curriculum Committee Task Force for CS1. *Communications of the ACM*; 1985; 27(10): 998-1001.

8. Koffman, E.; Stemple, D.; Wardle, C. Recommended curriculum for CS2, 1984–A report of the ACM Curriculum Committee Task Force for CS2. *Communications of the ACM*; 1985; 28(8): 815-818.

9. Mulder, Michael C.; Dalphin, John. Computer science program requirements and accreditation. *Computer*; 1984; 17(4): 30-35.

10. Ralston, A. The first course in computer science needs a mathematical corequisite. *Communications of the ACM*; 1984; 27(10): 1002-1005.

11. Ralston, A.; Young, G. S., eds. *The Future of College Mathematics*. New York: Springer-Verlag; 1983; ISBN 0-387-90813-7. Note: Proceedings of the 1982 "Williams Conference" that added impetus to the movement to add more discrete mathematics to the lower-division curriculum. The ideas presented in the book are far from exhausted.